the

BICYCLE DIARIES

To (ody)

Thank you for all your
hard work to make
our country more energy
efficient. Thank you
for your support, too!
Enjoy the adventure,

Dan

the

BICYCLE DIARIES

*my 21,000-mile ride
for the climate*

DAVID KROODSMA

RFC PRESS
San Francisco, California

ISBN 978-0-9914616-0-8

All photographs by David Kroodsma except where noted.
All maps by John Kelly.
Author photograph by Lindsey Fransen.

Cover and interior design by Erin Seaward-Hiatt.
Edited by Kirsten Janene-Nelson.

Published in the United States of America

10 9 8 7 6 5 4 3 2 1

*In memory of two family members
who made this trip possible:*

*Linda Dill Parker,
for her unquestioning support,*

&

*Rolf Bae,
for his enduring inspiration.*

Ride for Climate route,
November 2005 to September 2007

The legend of the map reads:

California to Tierra del Fuego:
16,000 miles by bike
November 2005 – March 2007

Massachusetts to California:
5,000 miles by bike
April 2007 – September 2007

o o o Boat segments

CONTENTS 🚲

ACKNOWLEDGMENTS

The following story is a true account of my seventeen-month bike trip through Latin America, as recounted from notes, photographs, video, and memory. The people in it are real, although a few names have been changed, and I claim any inaccuracies as my own. With the exception of the Honduras and El Salvador chapters, the narrative follows the order in which events occurred in real life. (I biked across part of Honduras, then El Salvador, and then part of Honduras again, but the narrative presents all of Honduras first.)

I wrote and revised this chronicle over a five-year period, during which numerous people, providing feedback and support, have helped shape the book before you. To start off, after I returned from my trip a number of friends and family hosted me while I pulled together the first draft. Thank you to Tom and Lauren Hunt, who hosted me the longest; you were great fun to live with. My parents let me stay with them for a few months as well, during which time they only occasionally suggested I get a real job. A few friends in Colorado also offered couches and inflatable mattresses: Judd Boomhower, Michele Minihane and Bryan Palimintier, Martha Roberts, and Kate and Eric Tribbett, and Mike Whitaker. And Wilma and Tony Bass and Abby Falik hosted me for a time in San Francisco. Thank you all!

A special thanks goes to my writers' group. In 2008, my friend Jill Patton (then Jill Redhage) got six friends together to meet once every other week to improve our writing. Little did Jill know this would mean the group's rotating members would have to read many versions of each of my chapters, including the early not-very-good drafts. The best way to improve one's work is through feedback, and honest feedback is the best. Thank you to Jill Patton, Adam Abeles, Rebecca Beyer, Amelia Hansen, Julia Jackson, Amber Johnson, Hannah Naughton, Sarah Shulman, and Eric Simons. Also, a number of friends read the entire text and gave suggestions, including Lex Beyer, Louis Eisenberg, Steph Greene, Tom Hunt, Ian Monroe, Matt Oglander, and Emma Wendt. My sister and brother-in-law, Kenda Kroodsma and Greg Oates, provided feedback and support for many chapters in the book. And thanks to my father, Donald Kroodsma, who provided a thoughtful edit after encouraging me to write this in the first place. And of course to my mother, Melissa Kroodsma, who helped me to finish the narrative.

Thank you to my scientific reviewers, who read some or all of the text: Dr. Chris Doughty, Dr. Alyson Kenward, Dr. Michael Mastrandrea, and Dr. Lisa Moore. Also, thank you to Latin American Studies professor Dr. Patrick Iber, who reviewed sections to make sure I didn't say anything too embarrassing about Latin America. Again, any scientific or factual errors in the text should be attributed to me, not to my reviewers.

I funded the editing and design of this book through Kickstarter, and a number of people made extremely generous contributions to help this book go to press. The most generous include:

ADVENTURERS—Josh Apte and Meghana Gadgil; Kate Arnold and Dave Ullman; the Bass family; Lukas Biewald; Marc Fantich; Lindsey Fransen; Taryn Fransen; Tim Frick; Jason and Laura Glickman; Steph Greene and Ryan Schuchard; Janet Grenzke; Apa and Adrian James; David Johnson; Bryan Keefer; Cyrus "Mangoman" Khambatta; Donald Kroodsma; Annie Maxwell and Adam Pike; Jesus Mendiola; Petra Mudie; Dennis Murphree; Paul and Kathleen Newman; Kim Nicholas; Benjamin Oates, Amanda Riley, and Zach Riley; Rick Pam; Charlie and Shirley Paterson; Jill and Ben Patton; Stan Pauli; Nicky Phear; Assaad W. Razzouk; Julia Silvis; Tinky Tina; Mariah Tinger and family; Amy and Eric Wagner; the Whitaker family; Joey Williams; Xavi

CONTINENT CROSSERS—Kris Fransen

CLIMATE CHAMPIONS—Lex Bayer; Melissa Kroodsma

My friend John Kelly, a former professional cyclist, created the maps. Editor Polly Rosenwaike read through an earlier version of the manuscript and provided an extensive list of suggestions that greatly improved the text. For the final heavy lifting I worked with editor Kirsten Janene-Nelson, who improved the flow on every page of this book and helped see me through countless details and decisions. If you find the narrative easy to read, you can credit her. Holly Cooper provided a thorough proofread, checking every word of this text and catching a number of items I would have never seen in a thousand years. My designer Erin Seaward-Hiatt developed both the cover and interior. And Michelle Lee designed the current form of RideforClimate.com.

I was fortunate to join an excellent team for Ride for Climate USA. Following my ride to Tierra del Fuego, I crossed the U.S. with Bill Bradlee, an advocate who organized this second leg of my journey. Countless people helped with outreach for this portion, but a key group of volunteers put in an incredible amount of work to help us reach a wider audience. Thank you to Melissa Borsting, Alan Duke, Angella Holmes, Jennifer Molfetta, Nicky Phear, and Martha Roberts. Thanks to them we shared our message with far more people than we could have reached on our own.

Thanks go to our sponsors: Tarptent provided Bill and me with great tents. I can't recommend Tarptent highly enough for lightweight travel— I was continually amazed that my spacious, comfortable shelter pack weighed less than two pounds. Chaco sent me some sandals when I was in Central America, and I wore them across much of the rest of Latin America. Greenerprinter provided Ride for Climate USA's printed material. Clif Bar provided us with an unending supply of Clif Bars for our journey across the U.S. by mailing us a box of bars to every other state. And Marmot supplied us with incredible clothing and sleeping bags—and then did so again when I went to Copenhagen in 2009, ensuring that I kept warm through the Danish winter.

And thank you to Lindsey, who read the manuscript more than once, has lived with me as I finished it, and has given me the next great adventure in life. (Lindsey and I will be spending much of 2014 riding across Asia.)

Finally, there are also the countless men, women, and children who made this trip possible—especially the more than one hundred people who let me stay in or behind their home over my two-year expedition, and the nearly forty fire stations who offered me a bed. And I was helped in so many other ways. More often than not I could not believe my luck in this journey—it often felt like interesting people hosted me just so I'd have a better story to share. Whether it was an ex-president in Guatemala, a petroleum engineer in Venezuela, or a firefighter in Peru, it was a pleasure—and an education—to meet everyone. It was all those I met on my travels who inspired me to write this book; for this I am forever in their debt.

And thank you, reader, for taking time to read this story. It is a joy to undertake such an adventure; it is a greater joy to share it. I hope you enjoy this book, and I hope it inspires you to see the world, to see the world differently, and to make a difference.

PROLOGUE
Crossing the Border

═══════════

I stopped pedaling and looked anxiously at the frontier. In front of me was the California-Mexico border, marked by a white tollbooth-like structure and a large sign that read Mexico in metal letters beneath the country's coat of arms: a large eagle perched on a cactus with a snake in its talons. It was an hour after sunrise on a Sunday in early December, and the light angled through the cool desert air, illuminating the quiet, dusty town of Tecate just beyond the tollbooth.

If I was trying to be inconspicuous, I was failing. Clad in arm and leg warmers, I essentially wore full-body Spandex, and I stood over six feet tall. My bike helmet had a large brim on the front and a cloth attached at the rear, both futile attempts to guard my pale skin from the sun. Between my outfit and my black touring bike with four red panniers, I wasn't going to blend in south of the border.

But, although I was conspicuous on the road, once I left it I was invisible. Unlike traveling by car, with a bike you can disappear off the highway at any point and set up camp, even just a hundred feet off the road, without a soul knowing where you've gone. Thieves can't rob you if they don't know you're there. It's the greatest sense of freedom; the world was my campground, my living room, my high-way. The night before, five miles north of Tecate, I'd easily found a place to sleep, as numerous foot trails led away from the highway, and a patch of desert ground had already been cleared—likely by people also wishing to keep hidden. I wondered how many illegal immigrants, how many of the people I saw working in California during my eight years in the state, had slept on similar patches of ground, staring at the stars.

At twenty-six years old, I was on the adventure I had dreamed of since I was twenty: starting from my home in California, I planned to ride south until I reached the southern tip of South America. I know this sounds a bit crazy. What sane person would camp along the Mexican border—the site of countless crimes and murders—let alone plan to ride alone across all of Latin America? Numerous countries I would travel through are besieged by the residual violence from recent civil wars; Colombia in particular endures an ongoing civil conflict, with nearly a third of the nation under rebel control. Moreover—except for in Chile, where I had studied for three months during college—I had no friends and almost no contacts south of the border. My Spanish was poor, and I had never traveled internationally by bike. Yet for some reason I chose to ignore all of these facts, and instead eagerly planned my route heedless of the risks. I decided I would ride solo across Central America, the Amazon, the Andes, and the wastelands of Patagonia, and I would reach the Earth's end in just under a year and a half. And not only did I want to bike this distance alone, I wanted to take the opportunity to raise awareness of climate change in the process—a topic I had spent the past four years studying. You might call me naïve. You might be right.

I had discovered bike touring six years earlier. After spending the summer after my sophomore year in college in the "biketopia" of Portland, Oregon, my friend Tom suggested that we "bike to school" in the largest sense of the phrase: riding the nine hundred miles from Portland to Stanford—and doing it in under two weeks. Pedaling a 1980s racing bike I had purchased that summer for $125, and carrying only a sleeping bag, tarp, and change of clothes, I biked with my friend along the rocky Oregon and California coastline. After just a few days, I was hooked.

While biking, no windshield protects you from the rain, heat, or wind, and there's no wall between you and the people along the road. In a car the scenery appears as if on television, framed by the windows. On a bike, you can practically touch everything you see. The world is right there, in 3-D. You also travel slowly enough that the person at the corner store can make eye contact and offer a greeting as you pass. And, off the highway, the bike becomes a prop for conversation, an excuse to talk. People lower their defenses and open their doors to cyclists in a way they don't for those traveling by car.

The trip to California took us nine days, riding as much as twelve hours a day. On that journey, we briefly met another cycle tourist who said he'd once taken off two years and biked to Patagonia. Our conversation was brief—I didn't even get his name—but the idea stuck with me. When we arrived at the palm tree–lined boulevards of Palo Alto, I didn't want to start my classes; I just wanted to keep heading south. I later read online about a couple different adventurers who had cycled from North America to Argentina, and nearly every one said it was the best thing they'd done in their entire lives. It seemed like the greatest way to see the world, the most invigorating and intimate way to experience the planet. And a trip like this would also be relatively cheap—the main costs just bike gear, food, and a return flight home. I calculated I could save enough cash for it in just two years—that is, once I finished school and started working. From that point on I hoarded my earnings, planned, and dreamed.

I wanted more, though, than just an adventure. I somehow wanted to make a difference in the world—even though, in retrospect, I had a very limited idea of what this world is like. I had spent my entire life in school or labs, and it would be a major leap to head from the classroom to the open road. This bike-tour dream needed some fine-tuning.

So in the meantime I continued with my junior year at Stanford University, studying physics, a major I had chosen just because I enjoyed it, not because I thought it would make me particularly employable. Two years later I earned my master's degree, in interdisciplinary environmental science. I happened to be interested in the most abstract of global problems—climate change. I found it fascinating how carbon dioxide cycles through the Earth's biosphere, lithosphere, atmosphere, and oceans—and how this odorless, invisible gas has the power to so greatly alter the world's climate.

I was researching this issue in the years before the award-winning documentary *An Inconvenient Truth* was released in 2006, before climate change emerged as part of the national discussion. Through reading scientific papers and attending lab group meetings, and working on an ecological experiment studying the effects of warmer temperatures, I had become convinced that climate change poses a real and significant threat to our civilization, and that too few people understood the magnitude of the challenge.

At some point I realized that these concerns dovetailed beautifully with my dream ride, so I decided to "ride to raise awareness of climate change." I would research the ways global warming was affecting the places I would be biking through—from California to Tierra del Fuego—and I'd encapsulate all I learned in a blog: www.rideforclimate.com. And as I traveled from region to region I'd share what I'd learned with each community, giving talks at schools and speaking with journalists about the very real threats climate change specifically posed to *their* environment.

In my head, it was a great idea; in practice, I had no idea how I would enact this plan in Latin America. I didn't even know how to say "carbon dioxide" in Spanish when I started. And, given that the U.S. pollutes far more than Central and South America, raising awareness in those less-polluting lands was a lower priority, Earth-wise. And never mind that the original goal of my journey was adventure and not activism, or that multi-year bicycle vacations won't solve a problem as enormous as climate change. I was just as ignorant about my activist goals as I was about how I would travel in Latin America.

In retrospect, my goal of raising awareness in others was obviously backward—it was my own understanding that changed the most. I saw that climate change is just one of many challenges facing humanity. In many ways, it's less urgent than ongoing conflicts in the Middle East, or than the daily poverty experienced by billions of people across the globe. And it isn't something that we should solve by restricting development: most people in the world rightfully want more, not less.

But all that being true doesn't alter the fact that climate change is real, and my trip gave me a vivid and personal sense of what is at risk—unique ecosystems, coastline settlements, and the livelihoods of countless people. While some of humanity will probably survive its consequences, I know we will regret what we will lose by not taking action.

At the border with Mexico, though, just a month into my ride, none of these thoughts crossed my mind. I didn't know what lay ahead or how it would affect me. I was just excited that the adventure had arrived, and that I was about to cross into Latin America. I didn't know my exact route, or whom I would meet, or how I'd raise awareness; I only knew that I'd ride south for months on end, until I ran out of road. I took a deep breath, inhaling the crisp December desert air, and pedaled toward the frontier.

PART I
CALIFORNIA AND MEXICO ڶ
An Adventure Discovered

California route

1 CALIFORNIA
The Journey Begins

California population: 38 million
Annual per capita GDP: $52,600
Annual per capita CO_2 emissions from fossil fuels: 9.9 tons

DAY 1
November 5, 2005

For two years, from 2003 to 2005, every weekday morning I biked to the end of my driveway and turned left, heading to work. I'd traverse four miles of low-traffic roads through Palo Alto's grid of one-story houses before arriving at the Carnegie Institution for Science, a small academic institute where I was a lab technician and researcher. I joked that one day, instead of turning left at my driveway, I'd turn right and head south. That day finally came on a sunny Saturday morning in early November, 2005.

I had planned to leave at 10 AM, but by noon I still wasn't ready. Though Argentina wouldn't mind the delay, I knew the eight friends I had convinced to bike the first twenty miles with me were less patient. While my friends milled around, talking with my housemates, I hurried to tape shut the boxes of things I was leaving behind. After selling my car and most of my possessions, I had reduced my old life to these few cardboard boxes stacked in my garage—a box of clothing, another of camping gear, and another of books (half textbooks from school)—plus a pair of skis and a racing bike. In theory, I'd want these things when I returned. But for now, I was content to maybe never see them again.

I had already packed all of the possessions I'd be riding with—sleeping bag, tent, stove, spare clothes, and bike tools. They fit tightly in four panniers hung on the front and rear racks of my bicycle, a black-framed touring bike with mountain bike–sized wheels. The way the sturdy two-legged kickstand lifted the rear wheel off the ground, my waiting bicycle looked like a bucking horse—except it was covered with decals. The night before, during my going-away party, I had encouraged my friends to decorate the frame with stickers. The front fender was now covered by a large image of the Virgen de Guadalupe, Mexico's patron saint. The bottom tube of the frame read ONE PLANET, ONE FUTURE, and spelled out on the rear fender were the words I BRAKE FOR NO ONE. The top tube read DEL FUEGO, my goal and final destination—Tierra del Fuego, an island off the southern tip of South America. Indeed, *del Fuego,* "of fire" in Spanish, became my bike's name, its reason for existence.

"Hurry up," Marshall yelled to me from the driveway, "we need you to stop global warming." My friends hadn't heard me talk about activism or outreach until only a few months earlier, though I'd talked about biking to the end of the Earth for years. The night before a former classmate had walked around with a digital camera and recorded friends' answers to the question, "What do you think of Dave's Ride for Climate?" Apart from my girlfriend, Ana, who defended the importance of my efforts, most used the opportunity to playfully mock me. Matt said he liked my journey because he could tell girls he knew someone who was biking to Argentina, and "just by knowing Dave, that makes me look better." Polly laughed and said, "I guess I'm a little unclear on the tie between global warming and personal recreation . . . I just want a better explanation of how recreating is saving the environment." Marshall added, "Actually, we figured out that, by biking so much, Dave will increase his metabolism to the point where, with the extra food he will eat, it's as if he's adding another small person to the planet. Do we really need that?"

I shared my friends' doubts. Why should anyone listen to me, a young researcher taking his message to the people? But, I figured, I was going to do the journey anyway; using it to raise awareness of a global problem was better than not raising awareness of anything—or so I told myself. And I had already gotten significant media attention about my trip, though sometimes it was hard to tell if the journalists were taking me seriously. The *San Jose Mercury News* had run a story, complete with a full-page map of my route, on the cover of the city section. The article began by saying "David Kroodsma—an Ultimate Frisbee–playing, climate-studying scientist—is out to improve the world, one bike ride at a time." While the story lauded my goals, it also made me out to be somewhat goofy, and poked fun at the fact that I didn't know how to talk about climate change in Spanish. But, then again, at least I was getting attention.

Finally ready, I rolled *del Fuego* out of the garage and addressed my friends while Mike and Teresa, two of my housemates, ran the video camera. We all yelled "Tierra del Fuego!" in unison, as if we were a high school cheerleading team.

My send-off committee and I mounted our bikes under a deep blue sky, pedaled out the driveway past a leftover Halloween jack-o'-lantern, and, finally, turned right.

―――――――――――

After twenty miles of rolling by Silicon Valley mansions and then climbing up Highway 9 into the steep, redwood-covered coastal mountains, all of my friends except for Ana bid their farewells. Twenty-five hundred feet above Palo Alto, at the junction with Skyline Boulevard, Marshall extracted two Coors Lights from his saddlebag and made me drink one with him to celebrate the departure. It felt surreal. I had been on rides with all of them, and this was like just one more Saturday excursion, except that now I was going to keep on going—for months.

Ana, equipped with camping gear, continued on with me, following the ridge of the coastal mountains south. She pedaled a steel-framed touring bike that I had found for her on Craigslist and then helped fix. Though she would soon turn back, we planned that she'd join me with her bike between Christmas and New Year's, so we could ride the southern Mexican coast together. We hadn't planned to still be "together" while I traveled, but we also hadn't planned to break up, which left us in a strange limbo.

First day of travel

Ana was tall, strong, and beautiful. We had been dating for half a year, interrupted by two breakups, and had spent many long conversations trying to figure out if we were compatible. Though we had different views on religion and God, we shared a love of outdoor travel, environmentalism, and laughter, and we genuinely wanted to understand one another.

We camped the first night on the soft needles beneath a grove of redwoods near Santa Cruz. In the morning—after a breakfast of oatmeal made with dried milk, heating the water with my little stove—we descended from the mountains to a coastal plain, biking by artichoke fields and through ocean haze that made the rolling farmland appear airbrushed. Forty miles later we arrived at the shoreline of Monterey, where the cold Pacific beats against rocky tide pools and harbor seals rest on thin beaches.

I'd spend the next day and a half in Monterey, visiting a local high school and then meeting with scientists at Hopkins Marine Station, a marine laboratory run by Stanford. Ana, though, would stay with me for only one more night—one spent on a lumpy foldout couch in the living room of a friend of hers. Then, early the next morning, we said our goodbyes, and Ana and her friend hitched a ride back to Palo Alto. The plan was for Ana to fly to Puerto Vallarta to meet me on Christmas Day, seven weeks and two thousand miles later. Until then I would call her every few days. But, for now, we shared a long embrace in parting, and I continued my journey.

DAY 4
November 8—Pacific Grove

The first stop of my awareness itinerary was just three miles away, at Pacific Grove High School. I rolled *del Fuego* into the front office just before classes started. I hadn't arranged this beforehand; my plan was just to visit a class, talking to the students about my journey and climate change. Since I couldn't schedule most visits in advance, this ad-hoc approach seemed my best option. I would set up countless presentations this way—simply by showing up and handing over my homemade business card to the receptionist. It was amazing how just a business card—which had only a map of my route, my website URL, and my ride's purpose ("To RAISE AWARENESS OF GLOBAL WARMING")—gave me authority in the eyes of teachers. It had worked at about fifteen schools already; before the trip I'd spoken about climate change to nearly a thousand Bay Area students. But when setting up a presentation "on the fly," as I was doing in Pacific Grove, the key was to arrive just before school began, or right after it ended. In Pacific Grove, I managed to speak to Ms. Smith, a science teacher, just before the first bell. After a quick introduction and two-minute discussion, she told me to come back to her environmental studies class at 10 AM.

Later that morning I faced a class of students clad in jeans and sweatshirts. They sat at long black tables, either slouched in their chairs or engrossed in conversation. After connecting my laptop to the projector and a short introduction from the teacher, I began my presentation.

"I'm biking from California to the southern tip of South America," I opened, describing my route and pointing to a map. "This distance, from here to Tierra del Fuego, is the same distance as from Spain to the east coast of China. It's like crossing the United States five times."

The students sat up. Two girls in the back who had been whispering to one another stopped; a boy looked up from doodling in his notebook.

I explained that I had come there hoping to get the students to think about their place on the planet. To start this off, we used the class as a human chart to represent the world's income distribution. As there were about twenty students in the class, I asked one student to stand; I then pointed out that only one out of every twenty people on Earth lives in the United States, and that on average we live on more than fifty dollars a day. I then had eight students stand, and explained they represented the one-third of the world's population who live on two dollars a day or less, and tried to describe what that might feel like.

"Because you live in Pacific Grove," I said, "you are many times wealthier than the average person in the world. If you were to follow me on my journey, or if you travel elsewhere, you'll see people who live very different lives than yours. I'm not saying you should feel guilty or proud about this—I'm just saying you should understand it. And you should also understand that we share the atmosphere with everyone across the planet."

The students didn't know how to respond to this information. Of course, I also hadn't yet biked across faraway lands, so I couldn't share meaningful stories of the different ways I'd witnessed people live. I didn't actually know what it meant to live on two dollars a day.

So instead I outlined the basics of climate change, explaining how we're currently performing a massive experiment on our planet, an experiment that will play out in these students' lifetimes. I described how, since the beginning of the Industrial Revolution, mostly through deforestation and the burning of fossil fuels, we've increased the amount of greenhouse gases in the atmosphere by more than one-third. The problem with the increase is that these greenhouse gases warm the planet, via methods that we've understood for over a century. With its normal processes of heating and cooling, the Earth is heated when it absorbs the sun's visible light, and then is cooled by giving off invisible infrared radiation. So the energy coming in from the sun is balanced by the energy going out, and the average climate doesn't warm or cool. But while greenhouse gases allow the sun's visible light to pass through to Earth, they trap and retain the infrared radiation that would normally cool the planet.

Because of our society's pollution, the Earth has warmed by close to 1 degree Celsius (just over 1 degree Fahrenheit) in the past century. And our pollution increases every year, largely because the world economy keeps growing. The problem is that carbon dioxide pollution is cumulative—it stays in the atmosphere for hundreds of years. "The carbon dioxide you put in the atmosphere today," I told the students, "will remain there until well after you die."

Students at Pacific Grove High School

"If we dramatically cut our pollution—and I mean *dramatically*— we might keep the Earth from warming more than 2 degrees Celsius (3.5 degrees Fahrenheit). And an Earth 2 degrees warmer would be noticeably different from today's Earth. Sea levels would rise due to melting ice sheets, some animals would go extinct due to shifting climates, and both storms and droughts would be more intense, because a warmer climate provides fuel for more extreme weather. But the thing is: right now we're not cutting our pollution—greenhouse gases are rapidly accumulating in the atmosphere. At our current rate of pollution, in your lifetime we'll likely warm the Earth by over 3 degrees Celsius [around 5 degrees Fahrenheit], and perhaps as much as 6 degrees Celsius [around 10 degrees Fahrenheit]. An Earth 6 degrees Celsius warmer would be unrecognizable. Forty percent of all plants and animals might go extinct, and all of the ice in Greenland and Antarctica would eventually melt, raising sea levels by two hundred feet."

I paused to let them reflect on this image, and then wound down my talk by showing maps of the countries I planned to bike through, discussing each region's vulnerabilities to global warming. I'd just finished collecting their email addresses to post updates about my trip when the bell rang, and my Pacific Grove students grabbed their books and backpacks and left the room. One kid yelled "good luck!" on his way out, and two others thanked me for the talk before running off to their next class.

I was left with just their email addresses. (High school students have the best email handles: "free2havefunx2," "rocktilldeath01," "sandwichwarrior.") Though I always hoped they'd follow my journey, I rarely heard from such students again, though every now and then I'd receive an email or a comment on my website from a student who had remembered me and was excited by how far I had biked. I allowed myself to imagine that at least some had remembered my short talk.

DAY 5
November 9—Riding Pains

From the Pacific Grove high school I charted a course east, toward California's Central Valley. Though a more direct and scenic route would have followed the coast south, I wanted to travel through the Central Valley, as its agriculture is at risk from climate change, and I wanted to see it for myself.

I rode thirty-eight miles until dusk, then camped hidden in a thicket behind a strawberry field. After inflating my pad and changing out of my bike shorts, I filled my small homemade stove with alcohol and cooked the first of countless pasta and cheese meals I would eat by myself, hidden in my tent, seated on my Therm-a-Rest. I found that I liked both the solitude and the independence of this arrangement. I especially liked the fact that, though I was so close to civilization, no one knew where I was.

Back on the road the following day, as I climbed east into the coastal mountains, redwoods gave way to drier, grass-covered hills, now brown from California's characteristic rain-free summer. My schedule for California was ambitious; I had appointments in Fresno and Los Angeles, and I had to ride long days—eight to nine hours, traveling at just ten miles an hour—to be on time. I planned to take a break once I reached Los Angeles, and fly to Florida to spend Thanksgiving with my family before crossing into Mexico. Having a fixed deadline kept the wheels turning, forcing me to log as much as ninety miles a day on a hundred-pound bike.

My bike was so heavy because it carted everything I needed for my new life. I had spent months carefully planning what I would carry: a single-walled tent, a sleeping bag designed for weather as cold as 40 degrees Fahrenheit, a lightweight inflatable pad, my little stove, and basic bike tools. I had two spare shirts and an extra biking jersey, as well as leg warmers and arm warmers, lightweight rain gear, and a fleece jacket. My six-liter water bag was useful for long stretches between towns, or for filling up before setting up camp off the road. I also had a small suite of electronics: a laptop (which I would mail to a friend once I reached Mexico), a PalmPilot with tiny folding keyboard (to replace the laptop south of the border), an iPod for music and to back up photos, extra memory cards, a host of chargers and cables, and a small tripod for my camera. I also brought a small electric razor powered by rechargeable AA batteries, which I'd use to look presentable to schools or the media. My mom joked that I was the "most high-tech homeless person." In a handlebar bag, which detached and could be carried like a purse, I kept my wallet, passport, camera, and a few granola bars for the road. I used Velcro to attach tent poles to the top tube of the bike, and a pair of sandals, my only shoes other than my bike shoes, to the top of the rear rack. I taped extra spokes—which I would need if my wheel tacoed in the middle of nowhere—to the frame between the pedals and the rear wheel, and attached a tiny bike pump between the seat and the pedals. All I would need fit on this two-wheeled mobile home, an RV without a windshield, roof, or walls.

I felt so free transporting everything I owned under my own power. But of course pedaling this mobile home was exhausting, especially at the beginning of the journey, largely because I didn't really train for the ride—I figured I'd get in shape on the trip. After half a day of riding out of Monterey, my legs were tight, my hands strained at the handlebars, and my rear was chafed from the seat, feeling as if fire ants had found their way into my bike shorts. I stopped at a gas station to refuel. In the United States, such convenience stores were unfortunately often the only place to buy food on country roads, which of course resulted in poor nutrition. I bought pasta and sauce for dinner, and bread, cheese, chocolate milk, and Doritos for snacks.

I camped that night behind a rise just off the road. It was probably private land, and I was likely trespassing, but, since I arrived after sunset and left at dawn, and my bike was hidden off the road, no one would see me. Exhaustion overpowered any feeling of loneliness. Out of cell phone range, I called no one.

DAY 6
November 10—California's Central Valley

Heading for Fresno County, the next morning I descended into the Central Valley, a wide plain of orchards and annual crops that extended to the eastern horizon. I passed a giant field of cotton, then one of corn, and another of grapes. California's Central Valley is an agricultural powerhouse. Although it accounts for less than 3 percent of the United States' farmland, the four-hundred-mile-long, sixty-mile-wide plain produces over half of the nation's fruits and vegetables. For a few specific crops, such as almonds, apricots, raisins, grapes, olives, pistachios, and walnuts, the valley is responsible for over 90 percent of the nation's harvest.

A few clouds dropped the autumn's chilly first rains. I put on my arm and leg warmers and kept warm by pedaling. People ask me what happens when it rains. I always answer, "I get wet."

I had actually planned my entire five-hundred-day journey following both the best bicycling routes and the climate calendar: balancing the best possible riding conditions over such a long distance and over such a long period. As such I timed my ride to avoid the rainy seasons as well as the extremes of winter and summer. But, while I avoided long stretches of heavy rain, there was no avoiding rain altogether.

During a break in this rain, I rolled to a stop next to a field where two-dozen Hispanic men were bent over between rows of bright red tomatoes. They followed a cart towed by a tractor that moved about one foot each minute, almost slow enough not to be noticeable. But the men's arms moved quickly between the tomato vines and cart, transferring the bounty in pace with the cart. Alongside the field, a middle-aged Anglo-looking man in jeans and leather boots stood next to a pickup truck. I biked over to ask for directions to Highway 180.

"It's really easy," he replied. "This is Belmont. Just follow this road until you reach Mendota, go south on 33, and then you'll hit 180. Can't miss it."

"Thanks. Is this your field?"

He proudly told me he'd been farming it for thirty years, explaining that he grew mostly tomatoes, but also had some almond orchards he'd been planting in the next field over. It took a few dozen workers, most of whom were migrants from Mexico, to work the fields. After this discussion, I decided to ask his opinion.

"I'm taking a survey of people as I ride," I said. "What do you think of global warming?"

The man's body straightened and his expression changed into a scowl. "I'll tell you what I know. It's entirely made up. There isn't a bit of evidence."

"What do you think of the science that shows that the Earth is warming?" I asked.

"Environmentalists are just making that up. They do this all the time. They also exaggerate the pollution from farms. You know why our air is bad here in the Central Valley? It's because of cars driving in San Francisco and San Jose, not us farmers."

"Well, yes, I somewhat agree," I replied. I know that the prevailing winds blow smog from the Bay Area into the Central Valley, where it is trapped and then combines with the pollution created by agriculture. "But smog is a different problem than greenhouse gases. Greenhouse gases are invisible. If we could see them we'd see that the atmosphere is getting a little darker each year."

"Look," he said, "in my experience, environmentalists just want to destroy the economy. They want higher fuel prices, which is the last thing I need. I guarantee global warming won't happen in my lifetime, or in yours."

"Has it gotten warmer during your time in the Central Valley?"

"No. It's gotten colder."

"Well, what would happen if water supplies were cut in half, or if temperatures were 10 degrees Fahrenheit warmer every day?"

"That won't happen," he said confidently.

In my soaking wet full-body Spandex and neon yellow jacket, I decided I wasn't going to convince him of anything other than my own lunacy. I thanked him for the directions and continued.

The encounter upset me, and the farmer's statements still burn in my memory, a reminder of the resistance environmentalism faces. I should have suggested that if we could figure out how to replace fossil fuels, we'd reduce our dependence on foreign oil. Or that if we invest in clean energy, we could build tractors that don't require as much gasoline. The key is innovation and progress, not sacrifice. At the time, though, I was thinking only of a study published by researchers in my former lab.

They had collaborated with other top scientists across the state to estimate the projected impacts of climate change on California, and the results were not good for agriculture. Regional climate projections are less reliable than global-scale projections, but they give us an idea of the future we may be headed toward. In California, for example, a combination of warmer temperatures, less rain, and less snowpack in the Sierra Nevada will likely reduce the yields of every major crop. In the later decades of the twenty-first century, the average July high temperatures in Fresno may be 115 degrees instead of 100.

With higher temperatures, dairy cattle and wine grapes, California's most valuable agricultural goods, would both see a drop in yields, which would be felt in the economy as well. Other valuable fruit and nut crops might be adaptable, but not without major investment. Warmer crops would also require more irrigation, a tall order given than climate models suggest precipitation might decrease by a third.

But even more troublesome than reduced rainfall could be the potential loss of mountain snowpack. While it almost never rains during California's summer, it rains frequently in the winter, causing deep snow to accumulate in the peaks of the Sierra Nevada. The snow acts like a frozen reservoir, storing water for months at a time and then releasing it slowly, hydrating California during the arid summer months. If current trends continue, by the end of the century 70 to 90 percent of the snowpack could disappear, leaving little water for irrigation. But, over the same period, California's population will continue to grow, thus further taxing the region's water demands.

What strikes me most about this future is how much the state's ecosystems will be reshaped. In the eight years I'd lived in California I had skied, biked, and hiked across its granite mountains, coastal redwood forests, lonely deserts, and oak-covered hills, discovering wonder in nearly every corner of the state. With climate change, this landscape will be forever changed. One study indicates that mist along the Pacific coast, which is essential for the survival of redwood trees, has been decreasing due to warmer temperatures. Similarly, warmer temperatures and dryer weather could cause the alpine forests in the Sierra Nevada to retreat, even to almost completely disappear, and would instead fill Yosemite with dry grasses and shrubs. The ecological impact would be enormous.

The same is true for ocean ecosystems. In Monterey, scientists at Stanford's Hopkins Marine Station had told me that the Monterey Bay had already been altered by climate change. A survey of species living in the tide pools was compared with a similar survey taken in the 1930s; it showed that marine species were already moving north in response to warming temperatures. A white-haired professor, Dr. George Somero, told me that many of the species living in tide pools are cold-blooded animals that can't adapt to even slightly warmer temperatures. "The local species of crabs, snails, and fish could disappear," he told me. "For those of us who love these places, it would be heartbreaking."

One could play the devil's advocate, pointing out that agriculture has changed so dramatically in the past few decades—with yields in California almost doubling in the past fifty years—that projecting out the next fifty years is a crapshoot. But even if we do genetically engineer crops to produce higher yields, extensive research has shown that warmer temperatures will almost certainly make yields lower than they would be without the warming. Or one could argue that we shouldn't worry about the scale of ecosystem change, largely because we've already dramatically altered the ecosystems by introducing invasive species and killing off predators like the grizzly bear—essentially that we lost "natural" California long ago. But, the fact that we've already spoiled the landscape is no excuse to continue.

After staying with a friend of a friend in Fresno, and having an inter-view with the *Fresno Bee,* I left town, biking south across Central Valley, through small towns whose residents seemed to speak only Spanish. Highway 33 was a low-traffic road that cut through a gauntlet of oil derricks bobbing up and down like a silent iron army marching in place. One night I spent camped in an almond orchard; on another I hid behind a vineyard in the coastal mountains. While the encounter with the farmer had dampened my enthusiasm for outreach, the adventure beyond that was everything I had hoped for. I was excited about my freedom, and about the fact that every day the Mexican border drew closer. The trip I had dreamed about for years was happening.

DAY 20
November 24—Thanksgiving

For Thanksgiving my parents and my sister and her family, who all still lived where I grew up in Amherst, Massachusetts, would be visiting my grandmother in Naples and then driving to Disney World. My parents had chipped in for my flight and Disney World tickets, so, when I reached Los Angeles after two weeks of riding and a couple presenta-tions in Ventura, I left *del Fuego* at a friend's apartment and boarded a plane to Florida.

Perhaps nothing could be more different from biking across the country and drawing attention to climate change than flying across the United States to go to Walt Disney World. Flying, of course, burns an absurd amount of fuel. The fuel needed to fly a plane full of passengers across the country is equivalent to what would be used if all the passen-gers drove the same distance, two people per car. Moreover, because of the physics of high-altitude clouds, the condensation trails behind planes further warm the planet. And to consume and pollute thus just to go to Disney World practically represents the excess of our society, and its lack of connection with the natural world.

Maybe it's hypocritical of me to say this, but I hope that solving climate change doesn't require us to give up the magic of flight, or even Disney World for that matter. Such things are wonders of the modern age, and environmentalism will get no traction if we declare we must give up these joys. In Disney World—accompanied by my nephews, two and five years old, and my niece, seven—I watched fireworks, laser shows, and a parade where Mickey and Minnie rode on a float with candy canes and a Christmas tree. I can't deny I enjoyed myself, and it felt like the right way to spend one of my last weeks in the U.S.

As much as I love my family, I was used to being apart from them for long stretches of time. But saying goodbye to Grandma, who had just turned ninety-one, was difficult. The last of my grandparents still living, she was also the one I had been closest to growing up. And as I had no plans to take another break during my journey, I wouldn't see her again for almost a year and a half. I hugged her close and promised her I'd visit as soon as I completed my adventure.

Having said my goodbyes, I boarded a plane and returned to California, where *del Fuego* patiently awaited me in Los Angeles.

DAY 24
November 28—Southern California

Del Fuego in Los Angeles—where it's "OK" to ride in the bus lane.

Los Angeles was a challenge for me; it felt as though it were built to exclude the bicycle. During my brief visit in the city—two and a half days—I planned to interview bicyclists about why they biked. But I found almost no fellow cyclists on the routes I chose, and those I did encounter told discouraging stories. One unshaven man told me he biked because his license was taken away, and then detailed the hospital bills he had accrued having been hit by a Mercedes, after which—to add insult to injury—a Lexus ran over his arm.

———————

It took more than a day to cross Los Angeles. I visited a school before leaving town, which was a welcome relief from traffic, but it also meant that I didn't have time to make it to my planned campground, and instead had to get a cheap hotel room near the intersection of two freeways. Hoping to avoid commuter traffic, the next day I left Los Angeles before dawn. My feeble red rear light blinked in the cold, dark air as thundering eighteen-wheelers roared past me on Highway 1. Before long *del Fuego* and I passed a large complex of smoking oil refineries. Tubes and pipes twisted around each other, climbing tall smoke stacks with flashing lights, like a dystopian futuristic city. I could almost hear the oil refineries taunting me, saying, *Get a car. Bikes have no place here!* I felt small and unwelcome. Only the slimmest streak of dawn on the horizon, behind the towers of steam-spewing pipes, gave me hope for the day.

Some sixty-five miles farther south and many hours later, I stopped at a convenience store near Marine Corps Base Camp Pendleton, just north of San Diego. I was desperate for sugar or some type of energy food, and my body ached with exhaustion. Though I was definitely getting more accustomed to riding, I still grew tired after six hours in the saddle. When I went to pay for an orange juice and energy bar, I reached into my pocket—*Where's my wallet?* I hurried out to *del Fuego,* where I had foolishly left it, but my travel wallet was gone—with my passport inside.

I was furious. *How could somebody do that?* Even more so, *How could I have been so stupid!* After half an hour spent in a frustrated daze, I biked to a campground ten miles away. I felt so naïve to have thought that traveling would be easy, that I'd be welcomed and respected wherever I went. If I was robbed at a convenience store in California, what hope did I have south of the border?

Luckily, friends in San Diego hosted me for a week while I waited for a new passport. After the initial frustration wore off, I was ultimately somewhat relieved to have this time. I was able to visit two schools, update my website, and, perhaps more important, research more about the road ahead. I was getting nervous about crossing into Mexico. In some ways, I was extremely well prepared. I had carefully chosen the gear I would carry, having honed my list of lightweight necessities. *Del Fuego,* durable and simple, was the perfect bicycle for the journey. Its steel frame would not give, its overbuilt mountain bike wheels would stay true, and its set of bike tools could repair most failures. And even though I hadn't perfected my own training for the ride, I felt healthy and strong—I knew I could withstand the journey. But, south of the border, I had almost no friends, and I didn't know if I could easily find places to stay or safely "free camp" off the side of the road. And one injury, one major robbery—of *del Fuego* especially—or one miscue could end in a second the trip I'd dreamed about for years.

During this waiting period I made my final phone calls before mailing my cell phone to a friend, as it wouldn't work in Mexico anyway. On one of these calls Ana and I spoke one last time. Losing my passport had put me a week behind schedule, and I was unlikely to make it to Puerto Vallarta in time to meet her there for Christmas. So, largely out of my urging, she decided to cancel her plan to join me.

But there was more to it than that. Though we had talked every few days or so since I'd left, our conversations had recently become shorter and more doubtful. My few weeks on the road had put physical and emotional distance between us, and I had come to feel the differences between us were too great to overcome. We—okay, mostly I—decided not to be in contact once I left California, and so this was goodbye.

I have ever since felt guilty and sad about this. I feel as if I betrayed something important by dropping communication and excluding her from the ride. She had given me much-needed support through all the planning and preparations—she partially owned this journey with me, as she helped me in some of the planning and shared in its beginning. Maybe we weren't meant to be, which is what I was focused on at the time, but I still feel like I could have given her a better goodbye, even if I don't know what that would have been.

So, both a bit heavier and a bit lighter, and with a new passport finally in hand, I mounted *del Fuego* and pedaled toward Mexico.

CALIFORNIA
November 5–December 10
Presentations: 7
Flat tires: 0
Trip odometer: 803 miles

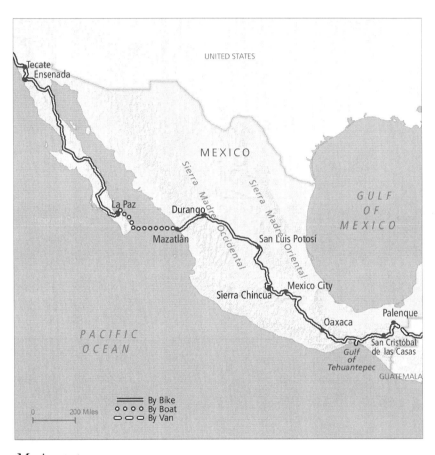

Mexico route

2 BAJA CALIFORNIA
Across the Desert

Mexico population: 121 million
Annual per capita GDP: $16,700
Annual per capita CO_2 emissions from fossil fuels: 3.8 tons

DAY 37
December 11—Tecate

No agent stopped me to check my passport when I biked through the gate, passing beneath the green, white, and red MEXICO sign. I pedaled down the palm tree–lined boulevard into the town of Tecate, and I felt invisible. I had entered Mexico and no one had noticed or cared. I also felt completely disoriented.

Spanish store signs stared down at me—LICORES LA LÍNEA, JOYERÍA PERFUMERÍA, CIRUJANO DENTISTA, ANILLOS DE GRADUACION—which forced me to dig for a long-dormant and limited Spanish vocabulary. I went to a Banamex ATM—I would rely on ATMs my entire journey—to withdraw two hundred dollars in Mexican pesos. These unfamiliar multicolored bills would have to last for ten days, as I was on a strict budget. If I could get by with twenty dollars a day, then I would have enough money when I reached Tierra del Fuego to purchase a return ticket home.

On my way across Tecate I passed a large green central square and a church with an image of the Virgen de Guadalupe prominently displayed. Beyond the main cluster of buildings, a network of dirt roads wove among small houses, many of which were simple concrete structures with metal roofs. There were no garages or driveways, though a few cars were parked along the streets. The community was evidence of the statistics I'd read: the average Mexican produces less than half as much carbon dioxide from fossil fuels as does the average Californian.

South of Tecate, the road cut through arid and unpopulated terrain that supported only shrubs, dry grass, and short trees. Trash lined the roadside: soda bottles, plastic bags tangled in bushes, tires, and aluminum cans. Most frightening were the crosses every mile or so, marking the places where the life of a driver, pedestrian, or cyclist had ended. I had heard and read that Mexican roads were dangerous due to reckless driving and poorly maintained roads; now I felt like ghosts along the road were echoing this warning. Though the highway I followed, Highway 3, had almost no traffic, I was still wary of each car that passed me.

───────

In the late afternoon, the road dropped me into a valley, where irrigation fed a series of vineyards between brown hills. So far on my journey I'd usually waited until about half an hour before sunset to find a campsite for the night, since that's generally how long it took to find an ideal hidden spot. But with the winter light fading as I rode past white-plastered cinder block settlements lining the road, I realized there were no places I could hide a tent. I kept riding, hoping to find a relatively unpopulated area. But when the sun dropped behind the hills in front of me, I still had no idea where to pitch my tent.

Having nervously steered *del Fuego* off the highway, I was following a dirt path, looking for a place to hide, when I was approached by a couple walking with their young son. The man, about my age, had a dark tan face with a mustache and wore a Philadelphia Eagles cap. The woman was dressed in black and wore glasses, her hair pulled back into a ponytail. I told them in halting Spanish, "I . . . am . . . traveling . . . by bicycle . . . I need . . . place . . . to put my tent *[mi carpa]* . . . for the night." I smiled and tried to look harmless.

My Spanish had once been pretty good; at one point I'd almost been able to read the first chapter of *Harry Potter* in Spanish. But it had been six long years since I'd spoken or read the language, not since my time in Chile, and now I stumbled over basic words. I hadn't found the time to brush up my skills before leaving Palo Alto, as I'd been busy with logistics—and my recently learned *"dioxido de carbón"* wasn't going to help me find a place to pitch my tent.

The man squeezed his thin eyebrows together. "*Carpa?*" he asked.

"Yes," I said, "for sleeping."

The man looked at his wife as if asking her permission, and then said, "You can stay with us. Follow us." His name was Julio, and his wife was Petra. Julio explained to me, by saying every word twice, that a camping tent is called a *casa de campaña*. "A *carpa*," he told me, "is what they use in the circus." No wonder he'd looked confused; he thought I wanted to put a circus tent in his yard.

"What do you . . . do here, Julio?" I asked.

"I work in the vineyard."

They led me down a dirt driveway to their home, which they shared with Julio's parents. "This is Geronimo, my father, and Marcelina, my mother." The one-story house smelled of dust, and had an uneven ceiling and concrete floors. The white paint on the walls was chipped around the windows, revealing the concrete underneath.

"Where do I . . . put my *casa de campaña?*" I asked.

"No, no. You can sleep here," Julio said, as he gestured me and *del Fuego* into a room with a small bed.

When I returned to the kitchen, Petra offered me garbanzo bean soup with chicken, fried tortillas, shredded lettuce, and my choice of Coca-Cola or Coca-Cola Light. After profusely thanking her, I asked, "How long have you lived here?"

"Twenty years ago, Geronimo built this house," she said, and I think she added that she had lived there five years. I asked if she'd been to California.

"No, but my brothers work there. Everyone goes there to work."

"Do you want to go?"

"Sometimes. There is no work here. Here you can only make five dollars in a day of work." Realizing my Spanish was poor, she spoke slowly.

"Are you married?" she asked.

"No."

"Girlfriend?"

"No." As I said the words, I realized how alone I was, and how much I was depending on that family to help me. It also felt like my old life was farther away than just a one-day bike ride across the border.

She looked at me quizzically and said something I didn't understand, but I gathered from her expression that it was along the lines of, "Of course, otherwise you couldn't travel like this."

After dinner we sat in the living room in front of the television. The small screen and poor reception made the show nearly unrecognizable, until at some point I realized it was *Forrest Gump*. I tried to follow along, but the Spanish subtitles were too small to read. I noticed the family seemed to watch more out of habit than for entertainment.

During a commercial break I gave Julio one of my business cards, which I had printed in English and Spanish, and told him in Spanish that I was biking "to give education . . . on global warming."

Geronimo, Julio's father, heard this and his eyes lit up. I handed him one of my cards, which he squinted at, his eyesight not good enough to read the small print. He said, "The winters are much warmer than they used to be."

César, Petra, Geronimo, Julio, and Marcelina on my first night in Mexico

"What do you do?" I asked him.

"I used to work in the vineyards," he said. "And once there was a huge fire . . ." and his eyes opened wide as he described what happened, but I understood little of it, except that he thought the fire was related to global warming. When the commercials ended Geronimo continued talking, moving his arms as he spoke, as if to describe how the fire moved across the hills, from one house to the next. I nodded and smiled, pretending to understand.

I was struck by how different Geronimo's response had been from that of the Central Valley farmer. Though the particular fire Geronimo described may have had nothing to do with climate change, global warming will certainly increase the number of wildfires. According to a study by scientists at Scripps and the U.S. Geological Survey, fires are already more common in the American West as a result of early springs and warmer summers. Wildfires in the fifteen years after 1986 were four times as frequent and burned six times as much acreage as in the previous fifteen years.

Most climate models predict that the deserts of Mexico and the southwestern U.S. will become even drier and hotter. A study in the journal *Science* examined the results of nineteen different climate models and predicted a permanent drought would be likely, and that the most severe of today's droughts will be considered "normal" by the end of the century. This could be hard for southwestern U.S. cities, where water shortages are already a problem, but it could also seriously challenge farming in Julio's valley.

I slept that night on a bed six inches too short for my long frame, my feet hanging uncomfortably off the end. But though the mattress was far more uneven than the foldout couch in Monterey, I was grateful for the hospitality, and to not be vulnerable and exposed in unfamiliar roadside terrain. In the morning I thanked the family and wrote down their address, which was only their surname and the town "Villa de Juarez."

DAY 38
December 12—The Baja Desert

I continued south from Villa de Juarez, riding to the coastal city of Ensenada and then following a secondary highway south. My plan was to bike the length of Baja California, a thousand-mile-long desert peninsula that hangs off Mexico like a tail into the Pacific. And then, when I reached La Paz at the tip, I would find a boat to the Mexican mainland. But after that, I had no plan. I would choose my route across central Mexico when I got there.

It would take three lonely weeks to get to La Paz. After visiting one school near Ensenada, where I quickly learned that I couldn't say a word in Spanish about climate change, I biked most of Baja without visiting a school, without a media interview, and without posting many blog entries. I did this partly because I wanted to cover distance, and partly because there just aren't many people to talk to in much of Baja. It felt good to take a break from the outreach portion of my journey and just ride for a time—though my mother wasn't too pleased, and sent me an angry email after just one week without any update. (Maintaining the website took almost an hour every other day, and keeping up with emails took just as long.) I apologized, but explained that I was busy biking across the desert.

Most nights I slept in my sleeping bag hidden off the road. For one stretch, I followed a dirt highway instead of the main highway, and had to carry three days of food and water across a washboard sandy surface. Most of the way across Baja, though, tiny towns every twenty or thirty miles offered stores with food and water.

Baja's lonely dirt Highway 3

It was magical to bike through the desert, riding beneath a blue sky that reached from one horizon to the other. The December days were short and warm, the nights long and cool. Each day I had fleeting conversations with people at small stores or along the roadside, which gradually awakened my dormant Spanish. I luckily got only one flat tire for the entire thousand miles; I quickly fixed it with my patch kit and pump. Layers of sunblock and dust accumulated on my skin, and I pedaled from sunrise to sunset, watching the sun chart a course across the sky. Each night I disappeared off the road, cleared a patch of sand, and prepared my bed for the night, either setting up my tent or, when it was warm out, using only my inflatable mattress to sleep under the stars.

My mind wandered as I rode. Sometimes I thought about my home in California, which I wouldn't see for more than a year, or I thought of Ana and our frustrating goodbye. When my blood sugar was high, I replayed great moments from epic movies in which I was the protagonist, charging downhill to fight the orcs or delivering the shot to destroy the Death Star. Or I sang the same song over and over in my head, replacing the lyrics with bicycling-relevant words ("You see I've been through the desert on a bike with no name, it felt good to get out of the rain"). When my blood sugar dropped, I pedaled along in a meditative, trance-like state.

The desert vegetation varied magnificently across the peninsula, providing visual evidence of my progress. On the northeast side, cacti replaced the bushes and short trees, and then, in some valleys, the rain shadow of the mountains allowed only the smallest sagebrush to grow in the rocky sand. In some basins, a strange plant taller than I am looked like a hand with many long alien fingers. Farther south, the cacti grew in size, reaching as tall as thirty feet. The world's tallest cactus, the cardón, is found along Baja's roads. I saw a few roadrunner birds sprint away through the bushes, and one morning when I awoke I saw a coyote run from my bike. At night I sometimes had to fend off large moths and other insects attracted by the glow of my stove.

The diversity of life in the different basins results from varying rainfall and temperatures. The northeastern shore of the Baja peninsula gets less than four inches of rain a year, while Ensenada, seventy miles away on the western shore, gets about fifteen. If rainfall decreases across the entire peninsula, the vegetation in each basin would change, perhaps migrating westward, so that the bushes and trees on the west coast—which can't migrate any farther west—would suffer the same fate as the farms Julio and his family cultivate: they would disappear. A cyclist riding here in fifty years would see a completely different landscape.

DAY 51
December 25—Loreto

I spent Christmas night alone in a hotel in Loreto, a small colonial town overrun by elderly gringo tourists. Through I managed to call my parents, my grandmother, and one good friend, I was enveloped by loneliness in the hotel. It was Christmas night, and I was in a room devoid of features or company, with no one to talk to and no Internet. I didn't wish to be in New England with my family, or California with my friends, but I did wish they could have joined me.

The next day I ran errands in town, and spent many hours at the Internet café updating my website. Then, when the time came for me to head back to the hotel, I suddenly remembered I might have a less lonely option for that night.

A month before departing, at dinner with friends in San Francisco, I had met Marcelo, who'd grown up in Argentina. When he was twenty-one he had gotten on his bike and left town with nothing more than a sleeping bag, a tent, and five hundred dollars. Over the next two years, with help from countless strangers, Marcelo biked from Argentina to Alaska.

When he learned I'd be biking through his homeland, he said, "You have to go to the *bomberos*!" *Bomberos* are firefighters. "They have extra beds at their stations, and they are all good people." Marcelo told me how, thanks to the *bomberos*, he'd had to pay for lodging only twice during his entire journey.

So I headed to the Loreto fire station and sheepishly asked a firefighter if I could set my tent nearby. "You can stay with us," he said. "We have a bed and, ah," looking at me, "a shower."

His name was Tecle. He was slightly overweight and unshaven, wore a red baseball cap, and had a gold front tooth. He gave me a tour of the two-room fire station, and introduced me to two other firefighters on duty.

"Our fire trucks were donated to us from New York," Tecle told me.

"Really?"

"All fire trucks come from the United States. A law in New York says that the trucks have to be retired after eleven years. So they donate them to Mexico." Eleven-year-old trucks in the United States are probably considered unsafe, or maybe too polluting. But in Mexico, they are better than no truck at all.

"What did you do before you were a firefighter?" I asked Tecle.

He replied with an answer I didn't understand. When I asked him to repeat it more slowly, he acted it out, and I realized that he was once a weightlifter. "But now I am fat." He laughed loudly and patted his belly.

I gave Tecle my card. "Ah yes," he said. "This is a serious problem, the heating of the Earth."

"What do you know about global warming?" I asked.

"Too much pollution," he said.

"Do you know anything else?"

"Well, it says here on your business card that your job is to explain it. So you tell me," he smiled. So I did.

Then he told me about his three children, and his life in Loreto, and we watched a Jackie Chan movie dubbed in Spanish. Tecle had a big laugh. He also snored.

In the morning, he told me about fire stations farther south. "Just tell them the firefighters of Loreto said you could stay there." A door had just been opened for me, and it revealed the road to Tierra del Fuego. The Loreto fire station would be the first of thirty-nine fire stations that offered me lodging in Latin America.

DAY 56
December 30—La Paz

On December 30, nearly two months after I had left my front door in California, I reached La Paz. Pleasantly located at the end of the Baja peninsula, its temperature a comfortable 70 degrees, the city of two hundred thousand felt like it was an island, as there were no other populated areas close by. The downtown curved along the edge of a placid bay, with palm trees rising from the tile boardwalk at the water's edge. Having now no more land to ride, I planned to make the two-hundred-mile trip to mainland Mexico by either hitching a ride on a private sailboat or paying seventy dollars for the ferry.

The *bomberos* welcomed me just as I'd been told they would, and they stayed up late with me, teaching pickup lines. Of course, hanging out at fire stations offered me few opportunities to practice these lines with actual women.

As I left the station the next morning, one of the firefighters said, "You should try to get on the news tonight. People should hear about your mission." Then he added, "And if you need directions or a ride, just give us a call." He wrote down the number 066, Mexico's equivalent to 911. "It's a free call from any pay phone."

———————

I thanked him and biked to the hilltop TV news building, which was covered with antennae and satellite dishes. My business card summoned a producer, who came out and asked "*¿Hablas español?*" My "*Sí*" was all he needed; he told me to return in thirty minutes for an interview. I laughed to myself. It was almost too easy to get on TV—just a business card made me appear official.

At my allotted time I went into the studio with *del Fuego* and took a seat by the desk of the anchorwoman, who had long, straight black hair and smooth skin. During the commercial before my interview she asked, "*¿Hablas español?*" Again I answered, confidently, "*Sí*."

When the commercial break ended she began speaking, quickly, the words rolling out of her mouth. I couldn't understand a word she said. When she stopped and looked at me, I realized it was my turn to speak. What should I say? I had never been on live television before. With the glaring lights on the set I couldn't see the cameras, but I knew they were out there, broadcasting me to La Paz. My pulse raced.

Politicians get away with shirking the question they'd actually been asked, instead answering the question they wish they'd been asked, so why couldn't I? And, besides, what other option did I have? After a few breathless moments, I said in Spanish, "I have been traveling for almost two months." To this brief "answer" she asked a few more questions, and I responded by muttering some hopefully intelligible statements about "*dioxido de carbón*" and the possibility of rising seas and stronger hurricanes, both of which would affect La Paz. I finished with my feel-good message. "We need to work together like a world team."

The crew at the television station seemed pleased with the broadcast. Afterward, I received big smiles, and the newscaster even invited me to join her family for dinner the following evening. Nonetheless, heading back to the firehouse, I felt uneasy. As a U.S. citizen, I was embarrassed to talk about global warming. The United States is one of the few nations that didn't ratify the Kyoto Protocol, an agreement aimed at reducing the pollution created by wealthy countries. While I do think the agreement was flawed in many ways, I object to the fact that we didn't try to renegotiate—let alone make any effort to reduce our own pollution. Instead of saying, "We can't do Kyoto, but we can do something else," we said, "We don't want to be involved." How can I talk about "working as a team" when my country has been the worst team player?

Earlier in December, when I crossed the border into Mexico, climate negotiations were taking place in Montreal, Canada. As George W. Bush was still president, the United States delegation did everything in its power to derail the talks, refusing to even informally discuss limits on pollution. And when a group of other countries declared they would limit their pollution whether or not the U.S. did, the head U.S. negotiator stormed out of the room in protest. I was deeply ashamed.

Even now, many years later, progress is still a huge challenge. The 2009 conference in Copenhagen demonstrated how difficult it is to get the world to agree on a climate deal. Switching away from fossil fuel use will take decades and enormous investment. How can countries fairly share this responsibility?

U.S. delegates, and especially U.S. senators (since the U.S. Senate has to ratify any treaty), often argue that, as China's yearly pollution now surpasses that of the United States, the U.S. shouldn't act until China does so first. But there are two problems with this stance—or this excuse. First, if we believe that people are equal, then we should be concerned about any pollution. And while China as a *country* pollutes more than the U.S., the average Chinese *individual* produces just one-fourth the carbon dioxide as the average American. Second, because carbon dioxide accumulates in the atmosphere, rates of cumulative pollution are a more significant concern than annual pollution.

Having had a head start as the most industrialized country of the twentieth century, the United States is responsible for nearly 30 percent of the carbon dioxide pollution currently in the atmosphere. By comparison, China is responsible for about 10 percent, and won't catch up with the United States for decades. If we include other wealthy countries—Japan, most of Europe, and a few others—the developed world is responsible for nearly three-quarters of the cumulative carbon dioxide from fossil fuels in the atmosphere.

But I don't like thinking about global warming in terms of our responsibility to reduce pollution. I prefer to think about our "ability" or "opportunity" to invest in new technologies *and* reduce pollution. But if anyone were to focus particularly on responsibility, they would point their fingers squarely at the United States. Anyone in the know in Mexico would have been right to ask me, "Why aren't you biking across the U.S.? Don't you know that all the countries you plan to bike across, combined, are responsible for less than 4 percent of the world's cumulative carbon dioxide from fossil fuels? While the U.S. is responsible for nearly 30 percent? And all of us joined the Kyoto Protocol; your country didn't. In fact, your country is the reason the problem isn't being solved."

Surprisingly, this situation never materialized; no one challenged the responsibility of my home country. But the truth is my primary goal wasn't to draw attention to climate change in Latin America—I wanted people in the U.S. to realize the global implications of our pollution. I hoped that the students I visited before I left would follow my trip and think a bit more about their place in the world. I had never really thought much about what I wanted people in Latin America to do. I had figured that would become clear as I traveled.

―――――――

The following day I visited the La Paz marina, where I met Sean and his girlfriend, Adrian, a semi-retired American couple in their late forties who were sailing to the Mexican mainland. Sean invited me onto their yacht, the thirty-eight-foot *Tiki Iti*, which they had sailed from San Francisco to La Paz. I excitedly told them I'd started in the Bay Area as well. They invited me to join them when they left for the mainland in a

Bomberos of La Paz—just call 066

few days. Thankfully, the *bomberos* had told me I could stay as long as I needed, so I had lodging until then.

I was eager for the next leg of my journey, through Central Mexico and Mexico City, where the majority of the country's people live. Having both successfully befriended a handful of firemen and survived live television, I was beginning to feel more confident with my Spanish. I had discovered Mexico's generosity, and making it through Baja's empty desert had helped me gain mental space from my old home. Little did I know that my journey up to that point had been merely a warm-up compared to what was to come.

BAJA CALIFORNIA
December 11–January 8
Presentations: 2
Flat tires: 2
Miles: 1,136
Trip odometer: 1,939 miles

3 CENTRAL MEXICO ᘛ
Into the Heart of Mexico

DAY 66
January 9, 2006—Mazatlán

On the sail across the Gulf of California, for the first time in my life I saw the sun rise over a horizon of ocean that stretched 360 degrees around the boat. This led to Sean, Adrian, and me disagreeing over who was more free—me on my bike, or them on their boat.

After leaving mostly unpopulated Baja, I felt fully in Mexico when we arrived in Mazatlán—as if the boat had transported me to a new world. Unlike Baja's parched climate, the Mazatlán air was humid and warm. The city has a large central square like many others I would see from then on, and the cobblestone streets I walked *del Fuego* along are centuries old. An outdoor market offered shrimp, mussels, crabs, and fish in white plastic buckets. A yellow stone cathedral stood in front of a large plaza filled with people, some wearing cowboy sombreros.

I found a man selling tacos and ordered two for twenty pesos— about two dollars. I tried to pay with a fifty-peso bill, but the man said "I don't have change for that" in a tone suggesting it was ridiculous to offer such a large bill. Still confused by the multicolored bills, I found a twenty-peso note.

"Spicy?" the man, appeased, asked as he prepared the taco.

"No, thanks."

The man laughed and said, "You're a gringo," and he gave me a taco with a near-perfect blend of onions, cilantro, tomato salsa, and chopped pork. While I ate he looked at my bike, and told me that the Virgen de Guadalupe, whose image was on my fender, would protect me as I crossed Mexico. Then he told me he used to work in California, where he made more in an hour washing cars than he now makes in a day selling tacos.

Perhaps I felt like I had finally arrived in Mexico, because my stomach revolted for the first time. I spent that night at the Mazatlán fire station throwing up in a toilet, my innards threatening to explode. The *bomberos*, as usual, were far too kind to me, and let me stay an extra day to rest my system.

I had not yet decided whether to follow the coast south to Guadalajara or ride into the mountains to Durango. Mexico has a giant central plateau with mountains lining its eastern and western edges. The Sierra Madre Occidental runs along the Pacific coast, and the Sierra Madre Oriental follows the eastern coast. Between these ranges, the dry, wide plateau of Central Mexico stands more than a mile high. In Mazatlán I was on the far western edge of this landmass, wondering whether to bike across it or around it. The decision was finally made when I emailed the operator of a small bike-touring company in Durango, who encouraged me to visit his city. So when I left Mazatlán I turned east, into the mountains.

―――――

The climb into the Sierra Madre Occidental challenged me more than any stretch of road yet, pushing my limits. The struggle began when my stomach ailment resurged, as if the poorly digested tacos were unwilling to let me enter Mexico's heart. I normally would have enjoyed the climb into the mountains. With green hills and thick moist air, the terrain was entirely different from Baja California. Deciduous trees lined the road, and a score of black vultures used updrafts to circle skyward. Hummingbirds pollinated budding flowers, and black-throated magpies, a blue jay–like bird with an impossibly long tail, cawed from branches along the road. Traffic reduced to a trickle. An occasional eighteen-wheeler passed, and truck drivers who waved to me were more common than those who didn't. But despite the scenery and pleasant traffic, all I noticed was the pain in my stomach and the weakness in my legs. I stopped twice, needing to expunge my stomach of breakfast and lunch. After only twenty miles, feeling dizzy and weak, I collapsed in exhaustion in a small town.

I spent three nights in that town, staying in a small stone hotel with maybe a dozen rooms, lying in bed while the braying of donkeys and the crowing of roosters kept me awake. The hotel was managed by a woman my age with long straight brown hair and a smooth face. She took pity on my illness and gave me a discount—ten dollars a night instead of twenty. She served me fresh yogurt and we chatted for hours about what it's like to manage a hotel or bike across Mexico. With my stomach mostly recovered, I hoped to steal a kiss or romantic evening from her before I left town. But she confessed she was married to the owner of the hotel, a fifty-five-year-old man she seemed to despise—a man who had gotten her pregnant eight years earlier, when she was fifteen. They now had two children, and she had to stay with him because he also employed her parents at the hotel. I felt lonely and sick, and I felt sick that I felt lonely. I was on the trip of a lifetime, and I was only sad because I hadn't found romantic companionship. She, on the other hand, was stuck in a life she didn't want, and still had many such years ahead of her.

My stomach recovered thanks to penicillin from the local doctor. The doctor, a former cyclist, was also eager to help with my journey—his wife kindly told me I could stay with her parents and siblings in Llano Grande, a town two days farther down the road. She wrote me a note of introduction, and I left town.

Though recovered digestively, I was still sapped of strength, and I struggled riding into the mountains. It was as if the illness had made my legs forget how to pedal, or that I had lost ten pounds of muscle. The steep mountain roads didn't help. As I slowly climbed higher the temperatures dropped, and short broadleaf trees gave way to tall evergreens. The road hugged a forested ridge that ran between deep, dry canyons, like a long green snake in a brown landscape. They call this route "The Spine of the Devil." The road climbed higher, reaching over 9,000 feet, and the canyons gave way to rolling hills. I camped, exhausted, in the evergreen forest.

I barely slept that night. Temperatures dropped well below freezing, and my homemade alcohol-burning stove performed poorly, barely producing a blue flame to warm my chicken-flavored ramen. Though the stove's fuel was available at pharmacies in every small town, sometimes it wasn't pure, which produced a weak flame. After a dinner of undercooked ramen, I retreated to the tent and slipped into my sleeping bag with all of my clothes on—arm and leg warmers, fleece, rain gear. As I had nothing left to use as a pillow, I stuffed an empty pannier under my head. My legs aching from the climb, I fell asleep, shivering.

DAY 76
January 19—Llano Grande

The next afternoon, exhausted from climbing and sleep deprivation, I arrived at Llano Grande, the high-altitude town where the family of the doctor's wife's lived. The town felt poor. There were no cars on the wide dirt paths between the houses. Families must have burned wood to keep their houses warm, because thick smoke filled the air. I had no address, just a letter with names. So I asked a few people along the road where the family lived, and the third man I spoke to pointed me to a house near the edge of town. A woman in her late fifties opened the door, read the note, and invited me in for the night.

I was simply amazed that she and her family were so generous, willing to open their doors based on a simple note. They didn't care that I was biking across the world or that I was talking about climate change. They just had a letter from their daughter, three towns away.

Eight family members, all grown adults, lived in three wood-and-plaster buildings with small windows. The father, who was probably also in his late fifties, was missing his left hand. He explained that he lost it cutting wood. I asked if most people there worked in forestry. "Yes," he replied, "that's the only way to make money." He made the minimum wage of about five dollars a day. His wife, a gray-haired woman who was blind in one eye, wore a sweatshirt that, like many clothes I saw in Mexico, appeared to be discarded from the U.S. She couldn't read the English written across her chest, so she asked me to translate. It read OFFICIAL HEARTTHROB.

I didn't know this at the time, but they were far wealthier than families I would meet later. Even so, though they owned a color television and had electricity, they cooked dinner, chicken soup, over a wood fire. I slept in the room of the doctor's wife who now lived in the town I'd come from, and I enjoyed having a bed, even if it was, as usual, too short. In the morning I joined the father and another daughter to pick up fresh corn tortillas, and then visited the church for a morning prayer. I asked the father if he knew other gringos. "I've talked to only one other in my life," he said. "A woman who hired us to help make charcoal." I suddenly realized that I was, as one of the few gringos they would meet, somewhat of an ambassador to the people in these small towns. The thought excited me, making me feel I no longer traveled as a tourist—I was truly off the beaten path. It also made me feel I now had some sort of responsibility.

I departed the next morning, legs still sore, and continued to follow the road upward. As I went on the landscape changed dramatically, and the evergreen forests gave way to the brown terrain of Mexico's plateau. The few surviving trees were small and shrub-like, and the sky was free of clouds. The road followed an arduous series of five-hundred-foot climbs and descents until finally, after a two-hour rolling descent, the city of Durango, with its nearly one million inhabitants, spread out across the dry plateau in front of me. I had made it across the first major mountain range of my journey. It would not be the last.

DAY 77
January 20—Durango

In Durango I sought out Walter, the owner of the small bicycle touring company whom I had emailed a week earlier. Coincidentally, his father had founded the American School of Durango; Walter took me to the school to inquire about my global warming presentation.

At the school I met a science teacher named Michael, who was from Illinois. He explained that most of the teachers there were from the U.S., and that the American School's classes are taught in English and follow a curriculum similar to U.S. schools. About a quarter of the students were foreigners; the remainder were the children of Durango's wealthier residents—tuition cost several thousand dollars per year. He recommended that I visit the many such schools throughout Latin America.

I talked to three of Michael's science classes, leaving out the part about wealth inequities, and the fact that the United States is responsible for a disproportionate amount of the pollution. Instead, I spoke mostly about what climate change would mean for the countries I visited.

In the last class I spoke to, a student wearing a HARLEM sweatshirt and Yankee cap asked, "Will it be like the movie *El Día Después de Mañana?*" In *The Day After Tomorrow,* global warming causes a sudden change in ocean currents, which triggers an instant ice age. "Will the U.S. have to evacuate to Mexico when it freezes over?" In the movie, the president makes a televised address from Mexico, thanking Mexicans for letting us enter their country as refugees. The students seemed to like that part. I told them that the movie was science fiction, and impossible. They were disappointed.

"But it does have a sliver of truth," I said. "Although the Earth's average temperature changes slowly, regional climate can change very rapidly." I explained how the layers of ice in Greenland and Antarctica have preserved a detailed climate history for the past few hundred thousand years. Over as little time as a few decades, average temperatures in a given region can change quickly, as much as 10 degrees Celsius, because of sudden shifts in ocean currents or wind patterns. Also, these records show that the climate during the past ten thousand years—all of recorded human history—has been far more stable than at any point in the past few hundred thousand years. If we perturb the climate, as we are doing, we could get swift and unexpected results.

"So, you're saying," the same student said, "that the U.S. might still have to be evacuated?"

I told him it was far more likely that Mexico and the U.S. Southwest would see a massive drought in the next few decades. The students looked much less enthused.

The region I was traveling through will likely become drier with climate change. Many places around the globe, mostly at higher latitudes or in some equatorial regions, are likely to become wetter. Will these rain patterns change quickly or gradually? It is hard to know. It is entirely possible that the onset of drought will be sudden.

Students at the American School in Durango

I don't know what such climate change would mean for the forests I had just biked through. The division of biomes—moist broadleaf trees near the coast, evergreen trees high in the mountains, and dry grasslands on the plateau—all have to do with average precipitation and temperatures. If the climate changes such that it is too dry for those evergreens, one likely scenario would be that the forest is gradually replaced by grassland. It's hard to say what that would mean for the family I had recently stayed with. The character of the landscape would be altered dramatically, and the entire town would need to find a new source of income.

DAY 84
January 27—San Luis Potosí

After a week's ride south of Durango—which included a talk at El Tec de Monterrey in Zacatecas—I arrived in San Luis Potosí, a sprawling city of new highways encircling a cluster of stone buildings in the historic downtown. There I received an enthusiastic welcome from Agustin, a soft-spoken bicycle advocate Walter had put me in touch with. Agustin led me around the city to give climate presentations at the local university, a middle school, and the town hall. I think he believed that I, as a scientist from California, would somehow convince the city government to expand the bike lanes. Agustin complained to me that the city built freeways but ignored the needs of cyclists, even though only 20 percent of the people in San Luis Potosí could afford cars. He emphasized how unfair it was to promote cars over bicycles. Riding through the city, I also felt this unfairness, but I felt conflicted as well. The government likely wanted to "develop," which meant more wealth and more cars. But as the traffic of this "progress" made it terrifying to bike or walk on most of the city's roads, the development was not necessarily a positive one for 80 percent of the city's residents.

Though I didn't usher in a new era of bicycles in San Luis Potosí, I was able to talk about climate change in Spanish for the first time, explaining the basics of the greenhouse effect as well as the storms and droughts that may accompany it. Of course, many people I spoke with confused climate change with either local air pollution—which is a huge, chronic problem in Mexico—or with the ozone layer. (It's an understandable confusion, given that some of the chemicals that damage the ozone layer are indeed greenhouse gases, but nonetheless they are two separate issues. Essentially, the ozone layer high in the atmosphere absorbs ultraviolet radiation from the sun before it can harm life at the surface. But, while the ozone layer has suffered some damage—mostly from chlorofluorocarbons—this damage is a very different problem from the climate change resulting from carbon dioxide pollution. Also, it currently doesn't pose as big of a threat, largely because clorofluorocarbon pollution is decreasing, while greenhouse gas pollution is increasing.) But still, I was impressed that every person I talked to seemed to agree that climate change was a serious problem, and they wanted to hear more about it. This was a marked contrast to the response I'd often gotten in the States.

In addition to arranging the three talks, Agustin also got the local paper to cover my project, and put me in touch with cycling advocates in Mexico City. I had once known no one in central Mexico. Now, because of a single email to one person in Durango, I had a network of individuals helping me advance to the country's center.

DAY 91
February 3—Sierra Chincua

After riding south another few days I entered the state of Michoacán. The terrain was less parched here—harvested fields of corn replaced the grazing land I'd been riding past. I spent the night at a fire station in the small town of Maravatío, and then started climbing into mountains rising high above Mexico's central plateau. Though I could have biked around these mountains—especially given my recent troubles with arduous climbs— I sought the monarch butterflies hibernating in their peaks. Every fall, a few hundred million monarchs migrate here from eastern North America, traveling thousands of miles. Congregating in a few groves of oyamel firs, they spend the winters hibernating in "super colonies" that require just the right conditions. The temperature must be cold enough for hibernation but not so cold they freeze; and the air must be dry and clear, as water can be deadly. In 2001, a single freak winter storm killed 70 percent of the monarchs—though the population had partially recovered since then.

A study published in the journal *Proceedings of the National Academy of Sciences* suggested that global warming would alter rain patterns in this region. Although the yearly total rain will probably decrease, more rain would fall during the colder months in sufficient quantity to kill wintering monarchs. As there are no other suitable winter refuges for the monarchs, especially because most of the surrounding forests are no longer standing, one possible result would be extinction.

Scientists debate how many species of plants and animals are at risk of being lost to climate change. One study estimated that in this century nearly 15 percent of species may go extinct. According to another article published in *Nature*, a 2-degree Celsius (3.5-degree Fahrenheit) increase in global temperatures—considered a relatively small amount of warming—would threaten 20 to 30 percent of all known species on the planet. Drawing on this study, the Intergovernmental Panel on Climate Change reported that if the warming is over 4 degrees Celsius (7 degrees Fahrenheit), we may lose 40 percent or more to extinction. Whatever the percentage, the message is clear: incredible biodiversity is at risk.

⸻

I climbed into the mountains, following a steep valley into thick evergreen trees. Clusters of houses lined a dirty stream. I passed a few dwellings with stacks of freshly cut lumber, and heard the buzz of electric saws. In theory, the nearby forests were protected, but in practice local residents and timber companies rarely follow the regulations.

The climb steepened. Using *del Fuego*'s lowest gears, my feet spun the pedals more rapidly than my wheels turned as I worked my way up the 7-percent grade. After three months of biking my legs were strong, and I had no fear of the hill, though I still traveled slowly. At an elevation of about 10,000 feet, a gravel road led me off the main highway to a large dirt parking lot full of cars. From there, a trail would lead me a mile through the forest to the butterfly colony. The sun hung low in the sky, though, so I would have to wait until the next day to see them.

At the trailhead, a trampled lawn hosted a circle of makeshift wooden booths where tourists crowded around and purchased souvenirs—postcards, magnets, and small plastic butterflies. A rusted metal trashcan overflowed with plastic bottles nearby. A child wearing sweatpants and a dirty white sweater approached me and extended his hand, saying in Spanish, "A peso. I want a soda." I gave him a banana.

Millions of monarchs in the Sierra Chincua Butterfly Reserve

I needed to camp, but I didn't want to stay near the parking lot. After talking to a park ranger and giving him my business card, I was permitted to ride down the path toward the butterflies and set up camp in the forest just a few hundred yards from where they were resting. Lying in my tent, I heard subwoofers from a nearby house beating out a rhythm. *How can a house be so close to the colony? This is supposed to be a protected area.* Climate change is clearly not the only threat facing the butterflies.

In the morning, beneath an evergreen forest and a perfectly clear cobalt sky, I walked to the fir grove. The trees looked like they had a strange bark disease, as all that was visible were the orange wings of millions of sleeping butterflies. As the day warmed, tourists ambled down the steep dusty trail. Only a tiny fraction of the butterflies awoke to spread their wings and fly, but this tiny fraction still accounted for tens of thousands of butterflies, making an animated collage of orange and blue. I passed the entire day sitting, watching, and marveling at this natural wonder.

DAY 93
February 5—Michoacán Countryside

After lunch I left the butterfly colony and biked nearly twenty miles, mostly downhill, by late afternoon. In reaching lower elevations, I again arrived in a landscape of harvested fields of corn. Interspersed between the fields stood the occasional concrete home. Looking for a place to camp as dusk fell one night, I stopped to talk to a woman standing outside one of these small houses. Her dog barked loudly.

Trying to sound friendly while yelling over the dog, I gave her my standard line, spoken in my gradually improving Spanish: "Hello. I'm traveling by bicycle and looking for a spot to put my tent for the night. Do you know where a good place would be?"

The woman's blank face suggested she didn't understand me. I smiled and thanked her and continued riding. I next inquired with a man, and then another; each said I could camp "farther down the road." Finally, a woman in black pants and a worn gray sweatshirt reading RALPH LAUREN offered the yard behind her house, introducing herself as Rene. Grateful for this help, I rolled *del Fuego* behind the two-story structure of unpainted concrete blocks and metal corrugated roof.

While I set up my tent on the grass, five puppies scampered around with two adult dogs, and chickens pecked at the ground around my bike.

I laughed and said to Rene, who stood nearby, "I like your chickens."

"Everyone here has chickens," she replied, surprised that they caught my interest. "Don't you have any chickens?"

"I've eaten many chickens," I told her. "But I haven't ever owned a live one." It made me feel somehow deprived.

"Well, we're having chicken for dinner. You can join us."

At Rene's dinner table I joined her thirteen-year-old daughter, Evarista, and twenty-year-old nephew, Javier. "My husband," Rene told me, "is currently working in Mexico City." She cooked our tortillas and chicken soup over a gas stove, not over a wood stove as I'd seen elsewhere in the countryside. She had electricity, too, which I was told had arrived in their town seven years earlier. With their two-story home and a gas stove, they were clearly wealthier than the third of the world's population that lives on two dollars a day, yet they were still extremely poor by U.S. standards.

The corn tortillas were thick and freshly baked. I asked if she purchased or made them. "No one here buys tortillas!" she told me, as if doing so would be sacrilegious. "I flatten them by hand, using corn from our land."

Javier had just spent a year working in Pennsylvania. "Why did you come back?" I asked.

"I wanted to see my family."

"That's not true," Rene said, somewhat scolding Javier. "You hurt your foot and couldn't keep working." Javier looked embarrassed, so I asked no more questions.

After dinner I retreated to my tent in the backyard. Lying down inside, ready to sleep, I heard the upstairs window open. A soft voice like a whispered yell called down. "Hello . . ." It was Evarista, who had barely said a word during dinner.

I sat up and looked through the small mesh window of my tent up to see her slight figure leaning out the window. "Hello," I replied. "Are you in school?" I asked.

"Yes. Next year I hope to go to high school."

"You will have to go into town for that, right?" The local village had a middle school, but the high school was in the larger town nearby.

"Yes." And then, "It will cost ten pesos a day. Actually, I'm not sure we will be able to afford it." Ten pesos was about one dollar.

"Then what will you do?"

"I'll stay here and work."

When she didn't continue, I asked, "Are indigenous languages spoken near here?" The Mexican government recognizes about sixty indigenous languages, and many villages still maintain languages other than Spanish.

"Yes, the village over there speaks Mazahua." *Over there* was the direction I had come from. Maybe that's why the first woman I had asked about camping hadn't responded.

"Do you learn that in school?"

"We have school in both languages," she said. "I know a little Mazahua."

"Say something in Mazahua."

"I know a song. Would you like to hear it?"

She launched into a short, pretty song I couldn't understand, her thin voice rising and falling in rhythm.

After I thanked her for her song she asked, "Do you know any stories?"

"Well, I have stories of biking across Mexico." I told her about the four fire stations I had stayed in, and how the firefighters of La Paz thought that their fire station was haunted by their former fire chief. I described how on New Year's Eve they blasted all their sirens at midnight.

"Do you have any stories?" I asked. No one had told me a bedtime story in quite some time.

Evarista told me about a farmer who was tempted by Satan. The Virgin Mary, though, appeared and revealed the devil's true identity, saving the man's soul and also his crop of maize. When she had finished, she said goodnight and closed her window.

―――――――

In the morning, Javier gave me a tour of the surrounding landscape, letting me ride the family's horse as he walked alongside and pointed out the various properties surrounding theirs. "What do you do here?" I asked.

"I work for my uncle, welding car parts." Javier explained that there was little work in the area and that everyone goes to the U.S. or Mexico City to work. "Do you think we make enough money here to pay for those windowpanes?" he asked, pointing at his aunt's glass windows. "No, that's earned elsewhere." Javier wore a leather jacket he'd bought in Pennsylvania. But I'd noticed that, despite his having worked in the U.S. for a year, he didn't try to speak to me in English.

"Everyone here grows their own food, right?"

"Yes, that's why people live here." He showed me their storehouse, a tiny concrete building that held a few months' supply of corn.

Javier and his family's horse

This region—central and southern Mexico—was in fact where, thousands of years before, the ancestors of the people now speaking Mazahua first cultivated maize. Corn's incredibly high productivity helped the native people build a civilization. When Hernán Cortés, the Spanish conquistador, arrived in 1519, he found a patchwork of different native city-states, many of which were ruled by the Aztecs. Cortés and a small Spanish army conquered and united these states under the Spanish crown, and modern Mexicans are a mix of the indigenous people and the conquistadors.

Although Mexicans have farmed for millennia, future farmers here will have to adjust time-worn practices to overcome significant climatic obstacles. Even though it was domesticated in tropical Mexico, corn, like other crops, does not respond well to warmer temperatures. Even a warming of just a few degrees will decrease yields in the tropics, which in Latin America stretches from Mexico to northern Chile. A study by a researcher at the Carnegie Institution for Science, published just before my journey began, showed that wheat yields in Mexico are lower in warmer years than cooler ones. Moreover, by the end of the century the average summer temperatures in Mexico could be warmer than the warmest summers ever recorded in the region, in conditions never before experienced. Another study by the Center for Global Development estimated that the country's average crop yields could decrease by 25 percent. Although some crops grown here respond favorably to higher levels of carbon dioxide, corn is not one of them.

As predicted by climate models (and as the U.S. Southwest already appears to be experiencing), this part of the world could also become drier. A likely outcome would be the expansion of the desert spanning Mexico and the southwestern U.S., as it would receive even less rain than it currently does. Less rain and warmer temperatures, which increase evaporation, can cause droughts. One study by the National Center for Atmospheric Research predicted that, in just a few decades, almost all of Mexico will have a climate considered "severe drought" by today's standards. The Western U.S. will also be in permanent drought, but one less severe than in Mexico. Today, the desert lies just to the north of where I was, and I could imagine it creeping southward, severely challenging farmers like Javier's family.

On the other hand, forty years ago activists warned that the world's population would outstrip our ability to feed ourselves decades ago. In 1968 *The Population Bomb,* a book by Dr. Paul R. Ehrlich, predicted environmental disaster and mass starvation was just around the corner. These predictions and warnings proved incorrect, however. The population grew more slowly than expected, and crop yields grew rapidly during the fifties, sixties, and seventies due to new crop varieties and industrial fertilizer. Agricultural technology saved us, and perhaps will save us again, with genetic engineering that allows us to grow food crops in the desert—or maybe even just greatly improved irrigation systems.

The extent of the effect on Central Mexico probably depends on how quickly its climate changes. If it is gradual, people may be able to adapt. But if it is rapid, with agriculture-crippling droughts in the next few decades, farmers like Rene and Javier would harvest little food. People whose families had cultivated the ground for generations would have no choice but to abandon their land.

CENTRAL MEXICO
January 9–February 6
Presentations: 7
Flat tires: 1
Miles: 966
Trip odometer: 2,905 miles

4 MEXICO CITY 🚲
A Landscape Transformed

DAY 95
February 7—Into the Smog

After riding two days from Rene, Javier, and Evarista's farm, I arrived at a pass overlooking Mexico City. Thick gray smog obscured the huge valley below. To my right and left, mountainsides with evergreen trees sloped steeply into the pollution. If the air had been clear, I would have seen the largest metropolitan area in the Western Hemisphere.

My heart beat quickly with anticipation. My plan had been to loop around the valley and avoid the metropolis, but as I approached it I couldn't resist the center of the country's economy and culture, the city where one out of every five Mexicans lives. After two months in Mexico, I'd improved my Spanish and my ability to travel by bike, and I felt ready for the challenge.

Different people had spoken about Mexico City, some with dread, others with excitement. Some had warned me of its dangers. A woman in Durango had said, "Whatever you do, don't go to the Buenos Aires neighborhood." A man in Toluca had said, "Never get in a street cab, only cabs that you call. The drivers might rob you." An American tourist told me he had been driving a new Toyota Tundra pickup truck in Mexico City when the police had stopped him. They claimed the truck's papers were not in order, but they'd overlook the violation if he paid them five hundred dollars. So, not surprisingly, the man had told me to avoid the police. But others encouraged me, with descriptions of its beautiful plazas and museums. Some noticed my sticker of the Virgen de Guadalupe and asked excitedly if I was going to the basilica dedicated to her. Others told me of the city's history, that Mexico City was once the center of the Aztec empire.

Following the road's switchbacks down the mountainside, *del Fuego* and I descended into the smog. With the help of gravity, we traveled almost as fast as the small cars and large, lumbering trucks following the narrow highway downward. I grabbed the brakes and leaned the heavy bike into the turns.

Buildings gradually emerged from the haze, revealing a residential district that crawled up the edge of the valley, with three-story concrete houses packed tightly together. Most of them were unpainted, although an occasional red, yellow, or blue one drew attention to itself. Laundry hung from clotheslines on many of the flat roofs, also adding color. Stretching out to the horizon, the rows of houses became smaller and smaller until they disappeared into the smog. The city seemed to go on forever, and I felt a shiver of excitement and fear.

Outskirts of Mexico City and smog

I pulled off the road and took out my new two-hundred-page *Guía Roji* map of Mexico City. I needed to ride across the city to the Portales neighborhood, to find the workplace of Judith, a woman I'd met briefly a week earlier in a town in the countryside. After just a short conversation she'd told me I could meet her in Mexico City and then stay with a friend of hers.

I followed Calzada San Esteban and crossed over the Periférico highway as I pedaled toward the city's center. As a slight downhill incline kept me moving, I easily kept up with the slow-moving traffic on the six-lane road. Then, as traffic increased and slowed even more, I was suddenly faster than most of the cars. Unlike in Los Angeles, where I was dwarfed by speeding SUVs, I felt like the king of the road, and was able to weave between the idling rows of small vehicles, most of which were the green-and-white Volkswagen Beetles that serve as the city's taxis. It was almost too much fun, and the chaos of Mexico City and the slower pace of traffic made *del Fuego* and me feel strangely welcome.

When traffic wasn't at a standstill, I rode in the center of the far right lane, demanding equal space on the road as if I were a car—as there was no shoulder for me to ride in. I watched my rear view mirror carefully, ready at any moment to bail to the sidewalk. Billboards lined the busy road: a large one for LG, the electronics company, another for Coppel, a department store chain, and one for Telmex, the Mexican telecommunications company. An advertisement for Hummers read LAND OF GIANTS, but fortunately few of them were on the road. I passed a Farmacias Similares, where a man dressed like Dr. Simi, the "mascot" of the pharmacy, danced outside the store. As the Mexican presidential elections would be held later that year, a billboard for Felipe Calderón read PARA QUE VIVIMOS MEJOR—"So we can live better."

Both my eyes and my chest hurt. I had been warned that Mexico City's polluted air would have this effect; the city's smog levels exceeded the World Health Organization's recommended limits two out of every three days. Smog is the result of a chemical reaction between sunlight and impurities in the exhaust of cars and factories. Because many of the city's vehicles are old and burn gasoline poorly, and because the mountains surrounding the city trap the pollution, Mexico City's smog is among the thickest in the Western Hemisphere. (It used to be among the world's worst until Chinese air pollution surpassed it in the past decade.)

The wide paved avenue, now Avenida Casa de la Moneda, gave way to the historic downtown, with colonial buildings on a grid of smaller cobblestone roads. I followed one of them and emerged at the central plaza, the Zócalo. A six-lane road surrounded the plaza, which was nearly two hundred yards across; pedestrians on the far side looked like ants. Unlike the squares of most towns I'd visited, the Zócalo had no greenery, no benches, and was completely paved with stone squares. Its only feature was a fifteen-story pole with a giant Mexican flag, whose green, white, and red contrasted with the gray of the plaza's stones and surrounding buildings. Circling the plaza just outside the six-lane road were stately buildings and a cathedral, all made of the same stones as the plaza.

This place was once the center of the Aztec capital Tenochtitlán. If I could travel back to the year 1500, I'd encounter a metropolis that was, in its day, one of the most populous in the world. To the right of where the cathedral now stands had been a pyramid that rose two hundred feet, and in front of it was the palace in which the Aztec ruler had lived. Leading away from the city center had been three wide streets, which had crossed the town and split into countless smaller roads lined with the adobe houses of Tenochtitlán's citizens. Perhaps most remarkably, the city had occupied an island in the center of a very shallow lake, Texcoco; canals cut through the town allowed navigation by canoe.

The lake and the ancient city are now gone. The Spanish leveled Tenochtitlán in the sixteenth century, destroying its buildings and decimating the population. A new city was built on its foundations, and Aztec pyramids and palaces were dismantled to construct the cathedral and other buildings. The lake had remained for a time, but, as the city grew over the centuries, the city government had built channels and tunnels to divert the water to adjacent basins, and constructed buildings on the former lakebed.

When I talk about climate change, I often say that we could completely reshape the world. The truth is, we've already transformed many places on the globe, especially our urban areas. This valley is now home to an enormous city instead of a giant lake. The population is a hundred times greater than it once was, and no Aztec would recognize it.

It's inevitable that we will continue to reshape the surface of the planet as the world's population and economy grows. In the future, I also might not recognize this city or the surrounding landscape. Perhaps the question, then, is not whether we want a completely different world, but what type of world do we want? Is it somewhere I would want to live?

Perhaps the cars of the future will run on hydrogen, their exhaust only water, thus freeing the air of smog. The mountains surrounding the city would reveal themselves, the sky would turn blue, and my eyes would not sting. Or maybe we'll continue to burn fossil fuels. Smog forms more readily in warmer climates, which means that climate change will worsen already polluted air.

I arrived at Judith's work before dark, and she enthusiastically welcomed me even though we had met only once. Leaving *del Fuego* at her office, we began a two-hour trip on subways and buses to the far side of the city and a community of gated houses far nicer than the concrete shacks I'd seen so far. I was to stay with a successful doctor who had an extra room, and then later in the week with another friend downtown.

Early the next morning Judith accompanied me back to the city's center, where I would later meet a reporter for an interview for the newspaper *Reforma*. The following day, on the front page of the City section of Mexico's most respected newspaper, an article accompanied a photo of *del Fuego* and me rolling through downtown Mexico City.

DAY 98
February 10—National Institute of Ecology

I spent a week in Mexico City, buoyed by my moment of fame. At one point a street vender recognized me and offered a free *torta*. When I gave a presentation at a middle school, students lined up and asked for my autograph. (I, of course, indulged them all.) A producer at the television station scheduled an interview, which aired on international television after I left town, and helped me set up a few school visits in southern Mexico. But, given that Mexicans pollute less than Americans, I still didn't know what my message to this country should be. At least I believed that I was inspiring the students I spoke to, and I was certainly getting more people to think about climate change. I was also having an enormous amount of fun.

Page 2 of City Section, *Reforma*, February 9, 2006: "Fight for the Planet by Bike"

Judith's downtown friend, Arturo, took it upon himself to show me different restaurants and bars, never allowing me to pay. Arturo was a journalist, and gave me the contact information for reporters in Colombia, Venezuela, and Peru.

I joined the Bicitekas, the bicycle advocacy group, for a nighttime "critical mass" ride through the City, in which two hundred cyclists took over the roads in a combination of recreation and activism. One of the Biciteka's members, Bernardo, invited me to stay at his house for two nights. He worked at a nonprofit that promotes bicycle transportation, so I joined him at work one day to learn about their efforts to build bike lanes in the city. While there I met Ricardo, a cycling consultant visiting from Colombia.

"You must come bike across my country," Ricardo told me. "You can stay with me in Bogotá."

"Is it safe? I'm not sure if I'm going to bike Colombia."

"If you stick to the main roads, you will be fine."

I told Ricardo I might see him in three of four months.

I couldn't believe my luck. So many people were helping me on my journey south.

———————

After visiting a number of museums and the basilica for the Virgen de Guadalupe, I bicycled to the National Institute of Ecology to meet with Mexican officials. The building was in a ten-story structure covered by reflective glass that mirrored the City's perpetually gray sky. After locking *del Fuego* near the entrance and asking a security guard to watch it, I offered the receptionist my business card and was directed to the office of Israel Monroy, who studies Mexico's policy on climate change. I found him in a small windowless room with two computers and many piles of papers. He wore a button-down yellow shirt and sported the ubiquitous Mexican mustache.

He recognized me immediately. "I saw you in the paper. Excellent job! Tell me about biking across Mexico!" He leaned back in his chair and rested his hands behind his head as I explained the joys of the road and the generosity of the *bomberos*. Then I asked him about his work.

He explained Mexico's position. "My current task is to compile reports on our emissions. The Kyoto Protocol, which we signed, requires us to do this. That's our country's only requirement," he laughed. Kyoto put limits on emissions in wealthier countries, but not on developing nations such as Mexico. Perhaps all of Latin America signed and ratified the treaty simply because the only requirement was to monitor emissions.

"What did you find?"

He handed me a tome labeled in Spanish: *Climate Change: A Vision from Mexico*. He flipped it open. "Here's a table showing Mexico's sources of greenhouse gases. Notice that our greenhouse gas emissions are different than a wealthy country, mostly because of this." According to the chart, Mexico's total carbon dioxide pollution is one-tenth the amount of that of the United States. *But* almost a third of it results from deforestation, while almost none of our pollution does. When forests are cleared for agriculture, carbon that was in the biomass of the forest and in the soil turns into carbon dioxide.

Deforestation accounts for roughly 20 percent of greenhouse gas pollution worldwide.

"Deforestation is mostly an issue of poverty," he said, and then he told me about a program to pay forests' property owners not to cut down their trees.

Since this discussion, I've looked more into the statistics on deforestation, and, echoing Israel's words, I've been surprised by how strongly deforestation correlates with poverty. Wealthy countries seem to be accumulating forests, while poor ones are deforesting. The history of the U.S. is the same. During the 1700s and 1800s, the forests in the eastern United States were cut down, so that farmers—whom we would consider poor subsistence growers by today's standards—could work the land. As the economy grew and farmland in the Midwest produced food more efficiently, people moved to the cities and most of the eastern farms were abandoned, which allowed the forests to return. Amazingly, the United States today has far more forested land now than it did in 1900. A similar story played out in Europe, and today the forests of Europe and North America—with a few exceptions—are expanding.

In his office, Israel also showed me statistics on Mexico's energy use, and compared them with that of the United States. The average Mexican uses about one-sixth the amount of electricity as the average American, and produces less than a quarter as much carbon dioxide from fossil fuels, though the latter is increasing. This story is fairly similar throughout the rest of Latin America, as Central and South American countries are both rapidly deforesting and increasing their fossil fuel use.

———————

After six days in Mexico City, I departed via secondary roads, biking southeast to avoid traffic. The mountains again emerged from the smog as I rode upward, my lungs now hurting less and less. Though I was filled with optimism, I didn't inflate the importance of my outreach, since the newspaper had mangled some of the details about climate change, even confusing the ozone layer and global warming. Nonetheless, more people in Mexico were aware of climate change than before.

Plus, I felt excited about my project and the road ahead. I had entered Mexico excited but naïve, and unsure if I'd meet people or make connections. I now had contacts in most major cities, and I had been far more successful than I expected in raising awareness about climate change through school visits and media appearances. It felt as if a power beyond me was charting my course. Maybe the Virgen de Guadalupe *was* assisting this tall, pasty-white gringo, guiding him safely from fire station to backyard to school to the cover of the national newspaper.

MEXICO CITY

February 7–13
Presentations: 4
Flat Tires: 0
Miles: 107
Trip Odometer: 3,012 miles

5 SOUTHERN MEXICO 🚲
Not All Roads Are Safe

DAY 109
February 21—Oaxaca

"Look at this gringo!" said Maria. "He should marry my daughter!"

"No, no," replied Margarita, "he should marry *my* daughter, Claudia!"

Margarita and Maria laughed, and the rest of us chuckled. *They're joking, right?* Claudia, who sat next to me at the table, reached under the table and squeezed my leg. She was my age and had long curly black hair, smooth olive skin, and a warm smile that gave me shivers of joy.

We sat at a round table with a floral tablecloth in Margarita's dining room. Her pleasant two-story house had a modern kitchen, high-speed Internet, and polished wooden furniture. Landscape paintings adorned the walls, as did a photo of Claudia's graduation from a private Mexican university. The only difference I sensed between this middle-class home and one in the U.S. were the bars on the windows and the multiple locks on the front door. Like most Latin American countries, Mexico has a highly unequal distribution of wealth: most of the country is poor, but as much as one-fifth of the population is more wealthy than the average United States' citizen.

"I'm not good enough for your daughters," I said, and laughed extra hard to make sure everyone was joking. "And I have to keep riding. A friend is flying to meet me in Belize in three weeks, so I can't stay around long enough to get married."

Margarita said, "Stay a few more days." Claudia smiled at me again. She was attractive, and she reminded me of what I'd given up for this ride. She was also twenty-five and unmarried. Most of the women my age I'd met already had children. She offered to give me a tour of the city later.

As Claudia's father, Enrique, passed me more of the *mole verde* and *frijoles,* I tried to change the topic by complimenting Margarita on the quality of her food. "See, you would be a good son-in-law," she responded.

I was here at the invitation of Margarita and Enrique, who had seen me on television and invited me via email to visit them in Oaxaca (pronounced "wa-HA-ka"), the capital of one of Mexico's southernmost states. The TV Azteca interview aired internationally on the show *Venga la Alegría.* I didn't see it, but they must have flashed my email address at the same moment the reporter asked if I had a girlfriend, because after that my mailbox was flooded with emails from all over the Western Hemisphere, mostly from teenage girls expressing their enthusiasm for "the climate." Many professed their support in all caps and a mix of Spanish and poor English. I felt a combination of amusement and pride. I also received a few invitations to give my talk at schools, including the school where Claudia taught, where I'd just spoken earlier that day.

That evening, Claudia took me on the tour through the narrow cobblestone streets of Oaxaca. I felt light and alive, and each smile we exchanged made me want to give up my journey and settle down. Claudia told me of her travels through Latin America, and how she would like to join me at some point south when school lets out in June. I reached out to her and we held hands as we walked down the road. I asked her if she thought that the students at her school would remember anything I said about climate change. She said, "Probably not, but I think you inspired them, and I don't think they will forget that." She took me to an overlook of the city and pointed out the landmarks, explaining how indigenous peoples built cities in the valley long before the Spaniards. Then, like high school students, we kissed and held each other before returning sheepishly to her parents' house.

A day and a half later, as I rode away from Oaxaca, I was filled with the deepest loneliness I'd felt since I had said goodbye in California. Claudia's family had invited me to stay longer, and I cursed my decision to continue on. But I had made a promise to myself that I'd complete this journey. I deserved no sympathy for my loneliness; I had decided to go on this quest and give up the comforts and joys of a stationary life.

DAY 112
February 24—Southern Oaxaca

South of Oaxaca, *del Fuego* and I rode through steep valleys lined by tall straight cacti that looked like cities of green skyscrapers. The hot temperatures of southern Mexico exhausted me, and the clear blue sky meant constant sun, forcing me to frequently reach for my water bottles. Sweat dripped down my forehead and arms then evaporated, leaving salty streaks. I rode in sandals to keep my feet cool. My bike shoes had broken near Mexico City, so I discarded them and replaced the fancy $80 clip-in pedals with $5 standard flat pedals. At midday, I found a shady spot beneath a low tree at the top of a hill, and collapsed on the dirt for an hour, resting before returning to *del Fuego.*

After three days like this, pedaling for an exhausting seven or eight hours each day, I descended to sea level, finally meeting the southern coast. Along this coast, a buffer of saltwater marshes, islands, and sounds separated the road from the Gulf of Tehuantepec and the Pacific. The air was even hotter than it had been inland.

After I rounded a corner the light wind became a fierce crosswind that nearly threw me to the ground. I leaned *del Fuego* into the wind and gripped the handlebars tightly to try to keep a straight course. Dust blew across the road, and plastic bags shook vigorously in the branches of bushes like the white pom-poms of cheerleaders.

Though I was on a tight schedule, I decided to take an extra day and detour the twenty miles to the ocean. I was curious how sea level rise might affect this area, and I wanted to take a rest day at the beach. Just before sunset I turned toward the sea, and the howling wind beat against my back, propelling me seaward. The road passed through a small community of wooden shacks, with villagers walking in the streets and children running barefoot. I didn't stop. The road turned to dirt, then to sand, and then to a thin trail that curved along the edge of a sound, where mangrove trees rose out of the sand and the moist swampy smell of the tidal flats blew by me. As the sun dropped through the shaking mangrove branches, the trail ended abruptly at a channel thirty feet wide.

Out in the water, a barefoot man in dirty cutoff pants, a tank top, and a blue Adidas baseball cap stood in a turquoise canoe, pushing his boat with a long branch. It was almost too dark to make out his face, but I could see that above his thin mustache his eyes showed surprise.

"I'm trying to make it to the ocean," I yelled to him over the wind. "Is it close?"

He called back to me. "Four kilometers from here, and you'd have to cross many channels."

"Then I need a place to camp. Is there anywhere around here you would recommend?"

"You can camp with us."

"You spend the night here?"

"Yes. That's the best time to catch the shrimp."

He pushed the boat to the shore and welcomed me to load the bike into the canoe. After he'd pushed the boat back again out to cross the channel, he introduced himself as Felipe and told me he shared the camp with his father and three brothers.

Once ashore we unloaded and walked to a sandy patch of ground surrounded by mangrove trees, their branches arching over us and shaking in the wind. A black cast-iron pot bubbled over a small fire. Nearby was a trash pile with an impressive collection of old and weathered two-liter soda bottles. A five-gallon jug of water sat on a wooden table. Blankets lay on the ground, and in a hammock hanging next to the fire lay an older man, the brothers' father. Felipe introduced him as Baltazar.

Baltazar had a gray mustache, a blue Yamaha baseball cap, a smooth tan face, and relaxed eyes. He'd been fishing for shrimp for forty years. "I started when I was fifteen," he told me.

"When was the last time you saw a gringo here?"

"You're the first one."

I couldn't believe it. It was as if I had passed through the looking glass and found a place that had never before been seen by my countrymen. I must have taken a wrong turn.

Baltazar confirmed I had taken a wrong turn—I had missed the road to the beach about five miles back. "The gringos always turn there," he said. He told me that he has another son who worked in Texas, and a daughter in Mexico City.

"Have you ever been to those places?" I asked.

"No, but I've been to Oaxaca twice." He was fifty-five years old, and the farthest he'd ever traveled was to a city a three-day bike ride away.

"Do you like shrimp fishing?" I asked.

"Yes, very much."

"Why?"

Baltazar shows off his catch of shrimp

"It pays well." He explained that on a good night they could make a hundred dollars per person. But, it turned out, that hadn't happened in years. Catches were down. I asked him why, and he said it was a result of overfishing.

After cooking a small pot of pasta, I laid my sleeping pad on the ground next to the blankets where his sons slept, just behind a wall of dead mangrove branches that served as a windbreak. The wind kept insects away, and I fell asleep to the sound of rustling mangrove trees as Baltazar and his sons set nets across the inlet.

The following morning they retrieved their nets. After removing plastic bags and trash, they found about five pounds of shrimp, which they could sell for about eight dollars. Baltazar held up a shrimp and smiled, showing no disappointment over the small catch.

Del Fuego and I then followed Baltazar, riding a single-speed bike rusted by the seawater, back to his small two-room wooden house. We shared an ice cream, which must have cost fifty cents. I'm not sure how he kept it cool, as he had no electricity. "Where should I put the plastic wrapper?" I asked.

"Just throw it anywhere you like," he said, waving his arm across his backyard. "The wind will blow it away." I put it in my pannier. Then he pointed me down the road to the ocean.

Again assisted by the wind, I biked the sandy road five miles to where a thin strip of sand met the Pacific. I stood for a while, watching the waves crash to the shore. As they crested the wind caught them and folded them back upon themselves, sending spray seaward. Nothing had been built along this windy stretch, and only a few fishermen were on the beach, readying boats with outboard motors to set nets in the ocean.

I asked a young man wearing jeans and a sideways baseball cap if he liked fishing.

"Yes, very much."

"Why?"

"We make about a hundred pesos a day. Working other jobs we would make about fifty pesos." I was struck by the fact that, whenever I asked people if they liked their jobs, they didn't tell me if a job was satisfying or not; they told me how much money they made.

The man offered me a ride in the boat, so I joined him and four of his fellow fishermen as they motored along the coast. The sea spray kept me cool despite the clear sky and tropical sun, and the long thin strip of undeveloped sand and mangroves stretched to each horizon. The wind howled as the fishermen set their nets, drowning out conversation.

———

As I left the coast, following a washboard dirt road northeast, I was struck again by the warmth and generosity of Baltazar and the other fishermen. They had not seen me on television or in the paper as the people in Mexico City had, but they still welcomed me to their camp, house, or boat. I suppose they would be friendly to all strangers, but probably especially so to a gringo who speaks their language and wanders off the beaten tourist path.

These villagers live by the ocean, and will be the first to notice the rising sea. As the Earth warms, Baltazar's community will be inundated with water from the melting ice sheets of Greenland and Antarctica. The ice that covers Antarctica, a continent one and a half times the size of the United States, averages seven thousand feet thick; the ice over Greenland, which is about one-seventh the size of Antarctica, is almost as thick. These two ice sheets cover only 3 percent of the world's surface, but if they melt sea levels will rise two hundred feet—the height of a twenty-story building.

The big unknowns are how much the Earth will warm, how much of this ice will melt, and how fast it will happen. To estimate how much ice will eventually melt, we can look at Earth's history. The Earth's climate has not been much warmer than it is today at any point in the past million years, and note that "modern humans" have been around only a few hundred thousand years at most. But, three million years ago the average global temperature was 2 to 3 degrees Celsius warmer (about 3.5 to 5 degrees Fahrenheit), and sea levels were between fifty and eighty feet higher. Thirty-five million years ago, the global average temperature was about 5 degrees Celsius (9 degrees Fahrenheit) warmer, sea levels were two hundred feet higher, and there was no ice at either pole. It is likely that we will warm the Earth by 3 degrees this century, and quite possibly by 5 degrees.

That doesn't mean sea levels will rise by two hundred feet right away. The ice is slow to melt, lagging considerably behind the Earth's rising temperatures, and it may take centuries or even millennia to melt all of it. In the short run, the best guess is that the sea level will rise between two and four feet by 2100, although there are a few scientists who believe it could rise far more. But in the long run, we will likely have many tens of feet of sea level rise.

As for just the next few decades: a rising ocean could swallow the mangrove forest I biked through. This would greatly affect Baltazar and the other fishermen of the community, as the mangroves' roots provide the habitat for shrimp and are the spawning grounds for many types of fish that later travel out to sea. The mangrove forest also plays a vital role for inland communities—as a storm break that dissipates the energy of waves when storms hit the coast. Without mangroves coasts would quickly erode, which would only amplify the effects of sea level rise.

Now, if the Earth were to warm slowly, sea levels would rise slowly, and the mangroves would adapt. As their knotted root systems trap sediment from streams, they could literally build up land as the water rises. But if the Earth warms more quickly, sea level rise could overwhelm the mangroves' ability to respond in this way. Alternatively, the mangroves might migrate inland to higher ground; but people live inland, which leaves no space for the mangroves unless towns are abandoned. One study estimated that if sea level were to increase by two feet this century—which is very likely—half of the world's mangroves would be lost.

I wonder, though, if Baltazar would choose to have electricity now if it meant his family would have to move in a few decades. But, then again, half of his children have already moved, looking for other places to live and make a living.

DAY 116
February 28—Citalapa to San Cristóbal de las Casas

From Mexico's southern coast, I planned to bike east across the isthmus of Central America, through northern Guatemala and into Belize. I left the coastal village before dawn, hoping to catch a pair of cyclists I had met a few weeks earlier.

On the road between Durango and San Luis Potosí, before I crossed into Mexico City, I had biked a few days with two American friends, Gregg and Brooks. Buddies from college and just a few years older than I, they had started biking in Alaska, and were also on their way to the tip of Patagonia. They were fundraising for the American Diabetes Association, as Brooks's mother had died of the disease.

Cyclists on long-distance journeys are members of an unofficial club. Every few weeks I'd run into another cyclist who was riding for several months. Every rider has a different style, and different goals. Gregg and Brooks rode elegant Co-Motion bikes with matching Ortlieb panniers and numerous gear sponsors (of which I was quite jealous). They stayed at hostels most nights, and carried high-quality DSLR cameras. In comparison I felt like a homeless vagabond. I had mismatched panniers, recently spray-painted brown to appear less valuable, and I spent most of my nights in a tent, on someone's floor, or at fire stations. None of that stopped us from becoming instant friends. No one else I met knew what it was like to bike across Mexico. I had enjoyed their companionship, and now I wanted to ride with them again. First, though, I'd have to catch them.

After half a day of riding, I asked a construction worker on the roadside if he'd seen two gringos on bikes.

"Yes, an hour ago they passed here," he said, describing them to me. A hitchhiker also confirmed that they passed. I was right on their tails.

Before dark I arrived in Citalapa, about fifty miles inland, and rode straight to the town's main plaza. I asked a man selling ice cream treats from a cooler if he'd seen my friends.

"They biked down that road, stopped at the *locutorio* on the corner, then went over to that corner by the church, took a few pictures, and headed out that road."

A man on a bench added, "That's exactly what they did."

I wondered how many Mexicans had such a vivid memory of me. Clearly riding through a small town on a loaded touring bike draws attention. We cyclists are vulnerable, with all our belongings strapped to our bikes. Bicycle touring is a statement of trust, a belief that people will respect us—and not follow us and steal from us.

Following the men's directions, I found Gregg and Brooks on a corner down the street.

"Look—it's El Payaso!" Gregg said. They had nicknamed me "the clown" in Mexico one day after seeing me with thick, white sunblock on my face and red salsa on my lips. "We thought you were ahead of us!"

We rode the next day together and climbed farther into the mountains. The air cooled as we rose, and my legs found strength as the temperature dropped. Gregg and Brooks told me about their time in Mexico City, Oaxaca, and on the southern coast.

"We spent a few days on the beach at a resort in Puerto Escondido," Gregg told me. "You should check out the pictures."

"Yeah," Brooks said, "Gregg is really proud of himself. He sat in his beach chair and took pictures of girls in bikinis with his telephoto lens."

Gregg smiled. "I'll show you later. How have you been doing with the *Mexicanas?* Brooks and I need to live vicariously through you." Both had girlfriends back home. I told them about the woman in Oaxaca, and that staying at a girl's parents' house, or staying at fire stations, are not the best ways to succeed romantically. They told me I was letting them down.

As we climbed, we talked about the area we were entering: Chiapas, Mexico's southernmost state. I had done embarrassingly little research on this part of the country, and Gregg explained to me what he knew. Many of the indigenous people in Chiapas, the Maya, were never conquered by the Spanish—many didn't speak the Spanish language—and thus were never fully incorporated into the Mexican state. As a result many complained of poor treatment by the government. On January 1, 1994, the day NAFTA went into effect, locals took matters into their own hands. Forming an alliance known as the Zapatista Uprising, militias overtook four municipal capitals and demanded more rights from the government, including work, land redistribution, and voting rights. The Mexican army retaliated, and in the fighting that ensued property was destroyed and over a hundred died. Though a sort of peace was negotiated a few years later, tensions have remained high ever since.

"Is it safe to bike here?" I asked Gregg as we rode.

"Yes. We'll need to be careful, but I think we'll be fine on this route."

The road became steeper. As a slow truck passed us, Gregg stood on his pedals and accelerated to almost fifteen miles an hour, the speed of the truck, and then reached out to the lumbering eighteen-wheeler and grabbed a rope strung along its side. "See you at the top!" he yelled. Brooks and I sprinted ahead and followed suit. Holding onto the ropes with one arm and our bikes with the other, we climbed the hill thanks to an engine not our own.

After two more climbs, the second of which we accomplished under our own power, we stopped at a roadside stand where a woman and a few children sold boiled corn on the cob. The corn tasted horrible—old and stale. The woman, who wore a colorful flower dress and had long, dark, braided hair, said to us in Spanish, "We are very poor." I tried to ask her how life is there, wanting to understand her experience.

"Dave," Brooks said to me, "I don't think she speaks Spanish."

I was embarrassed. The children, though, were fluent in Spanish, and I asked them what language they spoke.

"Tzotzil," they said.

I took out my notepad and asked the kids to write out basic Spanish words and their equivalent in Tzotzil. They obliged. "Hello" is *liote,* "goodbye" is *batan,* and "thank you" is *colabal.* I said *"Colabal, batan,"* to them in parting.

Indigenous child in San Cristóbal selling bracelets

Farther down the road, we passed a hunched-over man walking along the street with a small bundle of sticks on his shoulder. *"Liote!"* we called to him, and he stared at us in surprise. Later we yelled *"Liote!"* to a mother weaving at a roadside stand. She looked at us in shock and her children burst into laughter, either over our pronunciation or the sight of us trying to speak their language.

After climbing all day, we rode into San Cristóbal de las Casas, a tourist center with a colonial downtown. Temperatures were comfortably in the low 70s, as the city is over a mile above sea level. For the first time since Oaxaca I saw other foreign tourists. Young indigenous children in colorful dress accosted us and begged us to purchase bracelets and simple jewelry, or at least to give them a peso. They didn't look at me with the curious and friendly eyes I'd seen on the coast and in rural areas. Their eyes suggested I wasn't a person, but a traveling sack of money.

I couldn't look at these children without feeling deeply uncomfortable. The rural poor have land to grow their food; the urban poor must beg. I also felt sick when I saw the teenage boys in the courtyards asking to shine my shoes—especially because I was wearing sandals. Many of these boys were orphans. The disparity between what they have and what I have, between their options in life and my options in life, is enormous, and their begging eyes pierced me.

I paid one of the young girls to teach me how to say, "I do not want to buy that," as well as, "You are beautiful," in Tzotzil. Those phrases worked great in combination. The child smiled.

DAY 119
March 3—To Ocosingo, Agua Azul, and Palenque

The following day, after biking three hours out of San Cristóbal through cool evergreen-covered mountains, we stopped on the roadside to converse with a woman in Spanish. We learned that she lived in San Cristóbal but was in the country visiting her mother. She invited us into her mother's house, a one-room wooden hut with a dirt floor. She cooked handmade corn tortillas on a wood fire in the middle of the room. I greeted the mother, and was told, "She doesn't speak Spanish." I said, *"Colobal."* Her daughter said, "We don't speak Tzotzil here. The language here is Tzeltal. And another few kilometers down the road, they speak Chol."

Great, I thought, *so much for learning the language.* "Please tell her I said thank you." The woman translated and her mother smiled. She offered us tortillas with avocado, fresh off the fire. We enjoyed them immensely, despite the thick smoke in the room.

I found myself fixating on the fire. The primary energy source for people in the area is burning wood they cut from their land. This way of living, so removed from life in the U.S., is common elsewhere as well. Almost a third of the world's population—about two billion people—use biomass of some sort, wood or dung, as their principle form of heating or cooking. Unfortunately, these fires cause serious health problems. In fact, more than a million people die every year of diseases caused by indoor smoke.

Such open stoves also exacerbate climate change. The stoves release black carbon, small particulates of burned wood, into the atmosphere. The light absorbed by these particles contributes to climate change; it's estimated that black carbon is the third leading contributor to global warming after carbon dioxide and methane.

Plus, most of these stoves are woefully inefficient, and require about three times as much fuel as would a relatively inexpensive enclosed stove. If the two billion biomass users instead heated their food and warmed their homes with more efficient stoves, their health would improve *and* global warming would be slowed. The challenge, of course, is to distribute such stoves, make them affordable, and convince people to use them.

Continuing on our journey, Gregg, Brooks, and I descended the eastern slope of the Sierra Madre de Chiapas and arrived in a Mexico we had not yet seen. We were no longer in the land of cacti and brown mountainsides. Instead dense clouds filled the sky and the air was moist. Thick lush jungle had replaced evergreen forest, and large ferns arched over the road's edge. It felt like a different continent.

We occasionally saw men with machetes pacing the road. But it wasn't the machetes, which are used to clear jungle, that troubled me. These men were thinner, and their eyes were less friendly than those of people we'd met just thirty miles earlier. A boy ran out of a wood house, pointing his finger and yelling "Gringo!" as we passed, as if sounding a warning to those around him.

We stopped in the town of Ocosingo, where we shared a hotel room for the night. Though I'd wanted to ride another hour before camping, Gregg said he'd heard the road ahead wasn't safe at night, adding, "During the day it should be fine, though."

In the morning I left an hour earlier than Gregg and Brooks, as I hoped to travel farther that day. We planned to meet again in Central America.

Rain clouds passed overhead. For the first time since leaving California, I had to put the waterproof covers on my panniers. A warm rain soon soaked my clothes. I could see only a few feet into the thick green foliage on either side of the road, which was otherwise lined with few settlements. After an hour and a half of riding, I stopped at a three-walled wooden shack where a man sold potato chips and soda. I bought a Coke to raise my blood sugar. The man asked, "How much does your bike cost?"

I had been asked this question many times, but it felt uncomfortable in the jungle, where children yell "gringo" and men carry machetes.

"Two hundred dollars," I lied. The real cost was well over two thousand dollars. Belatedly I realized that two hundred dollars was still a lot of money—at least a month's pay. I decided to ride on and not stop until I reached a safe place.

That safe place was Agua Azul, a tourist stop where one can swim in the turquoise water of a small river. I had a long lunch break and then returned to the main road, expecting Gregg and Brooks to have passed me in the meantime. I asked a police officer if he'd seen them.

"Some people took their bags," he said.

"They were robbed?" I asked, in shock.

"Men with machetes stopped them when they were traveling slowly."

"Are they okay?"

"Yes, they're fine. They hitched a ride to Palenque in a truck."

"Is it safe to continue?"

"Sure," he said, "You'll be fine."

Weighing my options—and not trusting the officer—I jumped on *del Fuego* and started pedaling. It was thirty miles to Palenque and only three hours to dark. With the hills I could just make it. *Keep going!* The rolling terrain surrounding the road was a patchwork of jungle and land cleared for corn, and an occasional man with a machete stood on the roadside, walking to or from such a field. *Don't stop! Keep moving!* I told myself. I rode down the middle of the road, keeping well away from the edges. I felt ignorant, overly trusting, and terrified. I was incredibly naïve to think that I could bike every road safely.

Pedaling hard and dripping sweat, I stared down at the Virgen de Guadalupe. *Protect me,* I asked. *We've biked across Mexico together, and we've almost made it across your country.* She stared back at me, silent. As the road climbed, I slowed to five miles an hour, full of fear.

Outpacing the darkness, I rolled safely into Palenque just after sunset. At an Internet café I retrieved an email from Brooks giving the location of their hotel, and headed straight there. They were still dazed.

As Gregg described the place where they'd stopped for lunch I realized it was the shack where the man had asked about my bike. "After lunch, we rode out, and five men jumped out of the forest with machetes. But before they could do anything, a car rounded the corner, and they leaped back into the forest. So we kept riding, and tried to wave down cars to give us a ride, but no one stopped. Eventually, a police car drove by and we stopped him. We told him what happened, and he said he'd drive ahead to secure the road.

"We rode on, and three minutes later two more men leaped out with their machetes raised, yelling at us. They hit us with the handles of the machetes and told us to give them all our stuff. They looked nervous, which scared the crap out of me. One of them yanked on my rear pannier, jerking my bike up and down. I grabbed my tent and handed it to the man and then gave him one of my panniers. The other man got one of Brooks's panniers. Then we thought we heard a car and they ran into the forest.

"We biked ahead and found the same cop, who didn't seem to care about what had happened. Maybe he was involved. We convinced him to give us a ride to Palenque. We're done biking in Mexico. We're taking a bus to Guatemala. After that, we're not sure if we're going to bike or go home."

═══════════

We spent the next day in town and visiting nearby Maya ruins as we tried to decide how to continue traveling. Brooks was considering going home. "We've been in contact with about ten different long-distance cyclists," he told me. "Half of them have been robbed biking through Central America, some violently." Gregg said he might continue, but would first take time off from the bike.

Gregg and Brooks in Palenque, the day after the robbery

I had heard warnings that this last stretch of road in Mexico was unsafe. I had also felt the danger just in looking in the eyes of the people along the road. It was partly for that reason that I'd kept moving. I'd like to think I wasn't robbed because I was perceptive: at the place where I had felt uncomfortable and left, Gregg and Brooks had lingered, which gave the thieves time to prepare. A more likely explanation is that I was merely lucky. Also, it's possible I prepared the thieves for Gregg and Brooks, especially given my answer about the bike. After the thieves saw me slip away, they could have decided to rob the next cyclists.

That afternoon, Gregg and I biked to the police station in Palenque to ask about the road to Frontera Echeverria, on the border with Guatemala. I wanted to bike the highway if it was safe. A police officer in a beige uniform said, "During the day it is safe. At night it might not be."

"When was the last time someone was robbed there?" I asked him.

"I don't know."

"Can you find out?"

He walked into the station, then returned quickly. "Three days ago a pickup truck was stopped."

A pickup truck? "At what time of day?"

"Noon. But in the tourist vans you'll be okay."

"But I'm on a *bicycle,* this bicycle *right here.*"

"Oh, yes." The police officer looked over at *del Fuego* with interest. After a long silence, he said, "How much is that bike worth?"

"It was a gift. I have no idea."

"How about your watch?"

Gregg and I left in disbelief.

―――――――――

The next morning, we loaded our bikes onto the roof of a tourist van that would take us to the Guatemalan border. I felt defeated, sad to leave Mexico with anger and frustration when so much of the country had welcomed me with open doors, generosity, and friendship.

The jungle panned by at forty miles an hour as I thought over my time in Mexico. Had I changed? My views on poverty were different, or at least more nuanced. Most Mexicans are poor by U.S. standards. Some, like Baltazar, appear content with their lives, while others, like the beggars in San Cristóbal, struggle to survive. Nearly everyone, though, has far fewer options than I do in the United States. I understand why so many travel north to find employment.

But, overall, it was not the poverty that dominated my thoughts—it was the generosity. Nearly everyone I'd met, regardless of their wealth, had been willing to help me on my journey, providing food or a roof. They didn't ask me for anything in exchange, even though the gear I carried with me was probably worth more than many Mexicans will make in a year. As upsetting as it was, I knew this one robbery in southern Mexico was the exception, not the rule.

SOUTHERN MEXICO
February 14–March 5
Presentations: 3
Flat tires: 1
Miles: 945
Trip odometer: 3,957 miles

PART II
CENTRAL AMERICA 🚲
Hard Truths

═══════════════════

In Mexico I had discovered how I should, and should not, bike
across Latin America. I had mastered staying at fire stations or camping
in the backyards of campesinos. All I had to do was describe my journey—occasionally showing a newspaper clipping or business card—
and I'd be offered a place to stay. I had appeared on television and made
contacts that would help me the rest of my way south. I felt that if I
kept this up, those people responsive to my efforts would collectively
propel me along my route through Latin America. I also felt at least
some of the students I had spoken to had learned something—and that
they would remember the basic factors of climate change.

   If Mexico was my exciting discovery, Central America brought me
back to Earth. The novelty of traveling by bicycle was slowly wearing
off—I could no longer stay at a fire station for the first time, or visit my
first school in Latin America. At the same time, I gradually became
aware that I was riding through a region with more social challenges
than I'd yet experienced. By most standards, Central America is worse
off than Mexico, and many of the people I stayed with had stories of
past violence or severe economic struggles. Guatemala, Nicaragua, and
El Salvador are still recovering from the civil wars of the late twentieth century, and the region's murder rates are still among the world's
highest. A large portion of Central America's citizens live on less than
two dollars a day, despite the fact that each nation has a class of super-
wealthy people that support fancy malls in the capital cities.

I was also saddened because Central America is intensely vulnerable to rising seas, stronger storms, biodiversity loss, and agricultural failure. But, all the same, with the exception of devastating hurricanes, these enormous threats of climate change felt less immediate than the poverty and violence I witnessed.

I still felt as if supportive people were pushing me forward and the curve of the Earth was pulling my wheels south. I just felt less triumphant about it.

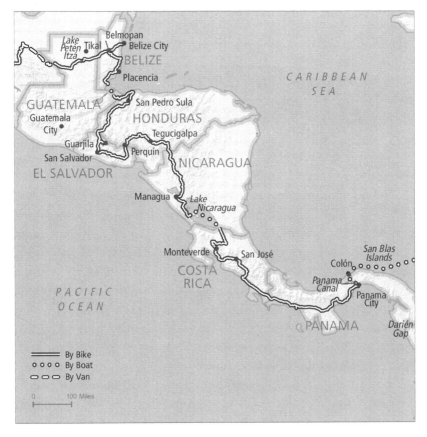

Central America route

6 GUATEMALA
Voices from the Past

Guatemala population: 15 million
Annual per capita GDP: $5,100
Annual per capita CO$_2$ emissions from fossil fuels: 0.8 ton

DAY 122
March 6—Across El Petén

Together we crossed the border between Mexico and Guatemala by taking a boat across a largely uninhabited stretch of the Usumacinta River—then Gregg, Brooks, and I parted ways. They had decided to suspend their bike trip, continuing instead in a small bus. I was still committed to completing my journey as conceived, but, before proceeding down the dirt highway leading away from the border crossing, I asked nearly everyone their thoughts about my safety—police officers, locals, the moneychanger. Although all gave me the same answer about the road ahead ("At day it is safe, at night it is not"), they were far more relaxed than the cops I had spoken to in Mexico, and few could remember a specific crime on that particular road. I realized I could endlessly research the road ahead, but ultimately I'd have to trust my gut on whether to pedal *del Fuego* onward.

I decided I would ride through only the northern portion of Guatemala, the country's jungle lowlands. My immediate goal—other than to reach Belize in time to meet a friend who would join me for a week—was to visit the Maya ruins of Tikal, a city that was abandoned more than a millennium ago. A number of studies suggest that a major drought contributed to the civilization's demise, and I wanted to see the remains of a city that may have met its end because of climate change.

As evening approached on the first night, after hours of pedaling along a lonely dirt highway, I saw a young girl, maybe ten years old, walk across the road. She carried a limp and half-grown chicken in her right hand. The chicken's feathers were thin and wilted, as if half of them had been plucked. The girl tossed the chicken into tall grass across the road from her house.

"Why are you throwing that chicken in the grass?"

"It's dead," she said.

"What killed it?"

"I did. It was sick, so we got rid of it," she said matter-of-factly, as if I were stupid for asking such a question. *Obviously one kills sick chickens— everyone knows this!* Since she hadn't also pointed at me and called me "gringo" like I'd recently experienced, I asked her about setting my tent somewhere nearby. She led me to her family's house, where I was told I'd be safe to camp. I chatted briefly with a few of the villagers, noting they didn't have the distrustful eyes I'd seen in Chiapas, and asked if any had crossed the border into Mexico. None had. I felt like I had traveled a great distance simply by crossing the border.

Campsite by a house in Guatemala

As I continued my journey the following day, jungle replaced grazing land on the roadside, but I was uninterested in my surroundings. I was getting sick. A scratchiness in my throat was developing into a cold, and each turn of the pedals became a challenge. Soon the jungle became a weary blur. But I wanted to make it to a hotel near the ruins of Tikal, so if I needed a few days to rest I'd be able to visit the ruins as I recovered. I kept riding.

Pedaling kept the endorphins flowing, which slightly mitigated my exhaustion. I told myself that light exercise is good for the immune system, though I suppose a fifty-mile ride on a hundred-pound bike doesn't count as light exercise.

DAY 123
March 7—Lago Petén Itzá

I arrived in Flores, the only Guatemalan city I would ride through, and immediately noticed small differences from what I'd seen in Mexico—the brands of gasoline, phone cards, and beer were all different, and the tortillas for sale were smaller and thicker. But I was only interested in making it to my destination. I continued for another ten miles through mostly uninhabited jungle until, as darkness fell, I arrived at a stretch of one-story hotels at the end of Lago Petén Itzá, a two-mile-wide, fifteen-mile-long lake surrounded by jungle. With swollen eyes and nose, I rented a hotel room for five dollars and collapsed in sweaty exhaustion next to the room's fan.

Previously when sick I've always wanted to be home, in my own bed, hopefully with friends or family to take care of me. But, lying on that hotel bed I realized that, though the sickness in my throat and the ache in my body had increased as I'd continued riding, instead of wishing to be home I'd thought only about the road ahead. I wondered if this meant I had now truly lost myself in the journey.

―――――――

At noon the following day I gathered my energy and walked to the more upscale hotel next door in search of an Internet café. In the lobby I encountered an unexpected sight: people I knew—a friend from childhood, Scott McNiven, whom I hadn't seen in years, with his girl-friend, Libby. They were on vacation, on their way to see the ruins of Tikal. We chatted about travel in Latin America and Scott's experiences in grad school. I briefly forgot I was sick.

Scott and Libby told me a team of scientists were staying at the lake, studying the past climate. I couldn't believe my luck. I then asked the receptionist about "the scientists" at the hotel; she told me they didn't get back until late, and went straight to bed. "You'll have to get up at 4:30 AM to see them at breakfast before they go back out on the lake," she said.

I stumbled back to my hotel, set my alarm for 4:30 AM, and spent the rest of the day napping or clearing phlegm from my throat.

In the morning's warm dark air, I walked into the dining room of the hotel next door. The room was empty except for three people sitting at a table with coffee and bread.

"Hi, are you the scientists?" I said.

The three looked up from their coffee with bags under their eyes. "Um, yes," said one of them, a skinny, well-tanned man with glasses and gray curly hair. He wore dirty jeans and a white T-shirt, and had an American accent. "Who are you?"

"I, ah, biked here from California." As I said this, the man tilted his head as if I were crazy. I quickly explained my project and my background in climate science and handed him my business card. He introduced himself as Mark Brenner, telling me he and his colleagues—all geology professors at the University of Florida—were drilling into the lake's bottom to research past climate. Though the other two then left to ready their equipment, Mark agreed to stay a few minutes to describe their work.

Every year, erosion from hillsides and runoff from streams provide the lake with new layers of sediment. By analyzing this sediment, the team could determine what the vegetation had been like in centuries past, as well as the relative amount of rainfall. Of course, that analysis entailed spending twelve hours at a time on a platform drilling three hundred feet into the lake's bottom.

While they could examine the climate twenty thousand years into the past, I was most interested in what happened during the "collapse" of the Maya civilization, a mere 1100 years ago. "I've heard that the civilization might have been wiped out by drought," I said. "Is this true?"

"Perhaps. There were massive droughts when these cities were abandoned." Mark explained that rainfall in the region was highly seasonal. From January to May the skies were bone-dry, but during the rest of the year it could rain heavily. The Maya in this region built reservoirs to survive the dry season, but the reservoirs could hold only about eighteen months' worth of water. "Between 760 and 910 AD there were a number of severe multi-year droughts."

Mark then explained that millions of people had lived in this region, and population densities before the collapse had reached many hundreds of people per square mile, which is denser than the populations in any country in Central America today. The forests had been cleared for agriculture in all directions, which led to seriously eroded soils. The bottom of some lakes have a twenty-foot-thick layer of mud that accumulated when the Maya cut the forest and the topsoil eroded and flowed into them. "Today, a thousand years later, the soil in the forests is thinner than it would have been if not for the Maya."

"Do you think soil erosion led to the end of the civilization?"

Mark shrugged and took a sip of his coffee. "It could have played a role." He then explained that when he'd first come there in the 1970s he'd researched soil erosion, only later looking at how climate could have affected the region. "We started with the idea that the Maya caused their own decline by destroying their topsoil. Then we said they were destroyed by natural climate change. And now I've read a recent paper that suggests rainfall decreased because deforestation affected the weather patterns and made rain less likely. So we've come back full circle to blaming the Maya." Mark said it was probably a combination of factors—the people had eroded their soils and populated the land to near capacity, and then the droughts had come. Climate change was probably not the silver bullet, but it likely played a role. "To me, the bigger question is how they managed to hang on so long given that the land must have been seriously degraded."

"I can see why people think of the Maya decline as a warning."

"Maybe. In some ways, I feel like I'm seeing the modern analog." Mark described how in the mid-sixties, about a decade before he started working there, only around thirty-five thousand people had lived in the entire Petén region of Guatemala, and untouched forest had stretched to the Belize border. Since then, the population has grown twenty-fold, to seven hundred fifty thousand. Although the current population is still far fewer than the millions who lived in the region during the ancient Maya times, history is, in some ways, repeating itself. "The forests are being cut down again, cleared for the first time in over a thousand years." Mark glanced at his watch, told me he had to head down to the boat, and said goodbye.

———————

At around 7 AM I set out to explore the ruins of Tikal. After paying an entrance fee and walking along a path through thick forest, I emerged in a grassy plaza where two pyramids of weathered limestone stood at either side. A chorus of sounds surrounded me—a rapid chirping of one bird, the longer caw of another, and the high-pitched buzz from insects. A black bird with a yellow tail and orange beak made a slurred beep, and along the grassy lawn of the plaza crawled a white-nosed coati, a brown and furry raccoon-like creature with a small head and long striped tail.

The pyramids, each about twice as tall as they were wide, had steep stone staircases climbing their sides. Atop each pyramid was a tall rectangular rock structure with a small door at its center, like a Cyclops head just above the treetops. I took a seat on the acropolis, a series of terraces facing the lawn between the pyramids. About ten feet away, a couple from the U.K. followed a guide, so I eavesdropped on his explanations.

"Tikal is called 'place of voices,'" he said, and then yelled, "SO . . . ALL . . . THIS . . . NORTHERN . . . ACROPOLIS . . . IS BUILT . . . LIKE . . . A GIANT . . . MEGAPHONE." As he yelled out we heard each word echo off the rock monuments. "The king would stand here and address the people gathered in the plaza." I somehow doubted the guide knew what actually happened—what we know is based on etchings in stone and archeologists' shovels, not papyrus, and certainly not live witnesses. But the echoing and the structures did suggest a past greatness, and I found myself imagining rows of soldiers and priests dressed in exotic wear, lining the plaza, listening to orders from their king, or perhaps performing one of the countless human sacrifices at the top of a pyramid.

Ruins of Tikal

Surrounding these buildings, other stone structures sprawled out through the jungle, including three other pyramids and the remains of a city that once supported many tens of thousands of people—with perhaps hundreds of thousands living in the immediately surrounding countryside. Similar abandoned Maya cities can be found throughout this jungle region, stretching from Mexico's Yucatán Peninsula to Honduras. The Maya were a civilization of millions of people, boasting technology that included a highly accurate calendar, a fully developed writing system, and even the concept of zero—which the Maya used hundreds of years in advance of the Old World. They also had sophisticated agriculture; where I now saw jungle, previously fields of corn and beans had stretched to the horizon.

Watching tourists wander around the plaza, I tried to imagine what Tikal must have looked like mid-abandonment. The city's government collapsed around 810 AD, which may have been in the middle of a nine-year drought. If that was the case, the city wouldn't have had enough water for its citizens, who may then have thought the kings could no longer summon the gods to create rain. Perhaps the lower classes revolted against their government. As the water supply dwindled, or perhaps when the crops failed, citizens would have fled the city. But where would these refugees have gone? Neighboring city-states would have been suffering the same fate. We do know that analyses of bones from this period suggest the Maya suffered from malnutrition and starvation.

According to studies of lake sediments—including research by Mark Brenner's team—during the collapse of the ancient Maya civilization, the rainy season, which typically lasts from June to December, never arrived. This was possibly because the Intertropical Convergence Zone (ITCZ)—a ring of moisture that circles the Earth near the equator and moves north and south with the seasons—didn't migrate as far north during those years, thus depriving Central America of its usual rains. This ITCZ anomaly may have been caused by natural variability in the Earth's climate, or it may have been because Maya deforestation affected the local weather. Regardless of the cause, the scale of climate change in this century is likely to be far greater than any variation experienced over the past few thousand years. Climate models predict that this region of northern Central America will become far drier in the twenty-first century, potentially even drier than during the times of the ancient Maya.

How will modern society's response compare with that of the Maya? In many ways, it seems easy to dismiss such a comparison. Tikal's residents were far more vulnerable to climate change than we are. They lacked metal tools, pack animals—even the wheel. A Maya farmer in the ninth century could produce enough food for only two people; today, a farmer using modern agricultural techniques can produce enough food for more than one hundred. The Maya lived much closer to subsistence level, and lacked modern medicine to aid those who fell ill. Nonetheless, it's hard not to feel warned while gazing across the remains of an extinct civilization. The Earth's population is growing; we may be approaching the carrying capacity of the planet. If we push the limit too far, perhaps we too will fall victim to climate change and ecological collapse. On the other hand, maybe such disasters could be averted by advances in technology.

I walked to the far end of Tikal and climbed another pyramid, pausing twice on the stairs to rest. I still felt sick and tired, but I tempered my thoughts considering how my discomfort was just a fraction of the worst Tikal's citizens probably experienced. Despite my cold, I was strong and healthy, and had long enjoyed an excellent diet. That night I would return to a hotel with clean water and cooked food; in times of their worst suffering the ancient Maya probably lacked both. When I finally reached the top, from my vantage point the other pyramids poked above the trees like giants wading up to their necks in foliage. Uniform jungle stretched to the horizon in all directions, and only the pyramid heads suggested that a great civilization once reigned here.

DAY 128
March 12—Pólvora

I spent the afternoon and following morning in my hotel bed, resting. By the next afternoon, I finally felt strong enough to bike again, and set out to make up for lost time. Having been delayed two full days, I now had to ride 135 miles in 24 hours in order to keep my appointment with my friend Dennis in Belize City. In order to visit Belize I'd also need to take a longer route through the region, adding at least a week of travel, but I knew it would be worth it.

I departed around 3 PM, following a low-traffic dirt highway, the only road connecting Guatemala and Belize. (The two countries have strained relations; Guatemala claims the British illegally stole the Belize territory from Guatemala, and that all of Belize should be considered Guatemala.) I had spent less than a week in Guatemala, seeing only the unpopulated jungle in an otherwise densely populated mountainous country. But such is the nature of bicycle touring; I experienced only a thin sliver of road, a one-dimensional transect across each country.

When night fell a few hours later, I stopped to ask three teenage girls along the side of the road if the highway was safe. They looked surprised by the question, replying, "of course it's safe," so I continued, deciding to ride at night. A waxing Moon hung overhead, providing me enough light to follow the middle of the road. No cars passed me.

About two hours later I arrived at a small town, Pólvora. Street-lights glowed above a stretch of palm-leaf-roofed wood buildings lining the dirt road. I bought two eggs and pasta at a store that had a roof but no walls. Inside, beneath the palm leaves, men sat around a table playing dominos, smoking, and drinking Gallo beer. When I saw a few teenage couples walking the street, I reminded myself it was a Friday night. I asked one of the men about continuing on to the border with Belize. "You can't cross at night," he replied. "The border is closed. Also, it's dangerous near the border."

"Is it safe here?"

"Yes, of course." I looked carefully at the people in the street and at the men playing dominos. I was checking for safety using the completely subjective, but I hoped intuitive, method of gauging suspicion in a person's gaze. I saw none. I asked the same man if they saw tourists here often; his reply of no made me feel safer. Seeing my concern, he added, "It used to be much more dangerous here five years ago, before they brought electricity and installed the streetlamps. Now there's less crime at night." *Electricity.* I looked around the town, realizing how transformed it must be compared to a decade ago, when every night was spent in the dark.

I befriended a couple who lived next to the store in a two-room building with thin wooden walls, dirt floors, electric lighting, and a gas stove. The man was drunk but harmless, and I set my tent next to his house, feeling protected by both the family and the glow of the streetlights. I cooked my pasta dinner, slept for four hours, and then departed two hours before dawn, racing to meet my friend in Belize.

GUATEMALA
March 6–11
Presentations: 0
Flat tires: 0
Miles: 130
Trip odometer: 4,087 miles

7 BELIZE
The Vulnerable Caribbean

Belize population: 320,000
Annual per capita GDP: $7,500
Annual per capita CO_2 emissions from fossil fuels: 1.4 tons

DAY 129
March 13—Belize City

"How many kilometers to the Belize airport?" I asked the border agent in English as he stamped my passport. The black man wore a blue button-down shirt and looked up at me with a smirk.

Speaking in English with a British accent, he said, "I don't know how many kilometers it is, but I can tell you that it's seventy miles."

Of course, I thought. *Miles. I remember those.* Belize used to be a British colony. I thanked the man and continued, glad to find that the road in Belize was paved.

Immediately seeing how little Belize has in common with its neighbors, I realized that crossing many borders through Central America was going to be disorienting. I would encounter different currencies and customs far more rapidly than I did crossing the Mexican landscape.

Belize gained its independence from the U.K. in 1981, and is culturally more similar to Caribbean island nations than to the rest of Central America: the majority of its population has African heritage, lives along the coast, and speaks a form of Creole English. The country is sometimes not even considered part of Central America per se.

Along Belize's coast lies the world's second-longest coral reef, bested in size only by the Great Barrier Reef of Australia. All reefs in the world are highly endangered by climate change, so I wanted to experience one while I still could. My plan was to join my friend Dennis, who was flying into the airport, and spend a few days scuba diving along Belize's coast.

———

"This reminds me of rural Connecticut," complained Dennis, who was originally from Texas, as we assembled his bike on a grassy lawn by the lone terminal of the Belize City airport. Dennis explained how, looking from the airplane window, he'd seen rolling green hills extending to the horizon, and he felt like he'd never left the States. "I mean, they even speak English here." Dennis, who usually wore a mischievous grin, instead looked frustrated. He was tired from travel, he wanted to practice his Spanish, and he somehow hadn't processed that he'd agreed to join me in the only non-Spanish-speaking country in the region.

"Dennis, stop complaining. I gave up marriage with a beautiful girl in Oaxaca to bike with you."

"Wait a minute, this is *my only* two weeks of vacation."

"You're a graduate student. You know, graduate student, *gradual* student. Taking vacation only means you'll graduate two weeks later. What's the big deal?"

"You also promised me forty-mile days on flat roads along the Caribbean. Now you're telling me we're biking sixty miles inland."

"Plans change." I had been invited to speak at an inland university, and also to meet with officials in the country's capital.

We rode from the airport into Belize City, where we got a hotel room for the night and bought sandwiches at a Subway. The following day, before leaving town, we visited the local television station and appeared on the national news—they also interviewed Dennis, simply because he was also on a bike. It felt odd to do such an interview in English instead of Spanish—it made it much easier to describe my journey and what climate change means for Belize. Dennis, of course, noted that because just over three hundred thousand people live in Belize, fewer people heard our message than if it had been broadcast in rural Connecticut.

We then rode the highway I had come in by the day before. Closer to Belize City, short brush grew in sandy soils along the road; as we rode inland, taller lush trees replaced the brush. Despite the fact that the road links the country's largest city, Belize City, with its capital, Belmopan, it was nearly deserted—a car passed us only once every five minutes. The countryside appeared devoid of people, and had even fewer settlements than the Guatemala lowlands I had passed through the week before.

After a day of riding we rolled into Belmopan, a town of only seven thousand inhabitants. Its central market was two rows of wooden booths in a hundred-yard-wide parking lot. The booths sold a range of produce, from familiar to unidentifiable—potatoes, beans, broccoli, bananas, lemons, pineapples, kiwis, then large green melons, a red pear-like fruit, a large cherry-like item, and a large yellow tomato-like fruit. Another stand sold Belize's two most popular meals: beans and rice *and* rice and beans, which, remarkably, were different meals, with different ratios of beans and rice. We bought both, and split the meals for variety.

Dennis and I wheeled our bikes across a lawn to a four-story government building. Leaving our bikes in the lobby, we proceeded to the office of Anthony Deyal, an official working on a United Nations project called Mainstreaming Adaptation to Climate Change (MACC).

Anthony, a Hispanic man probably in his forties, stared out at us through thick glasses. "The basic idea of the project is to prepare Caribbean nations for global warming," he said. "Island nations of the world produce less than 2 percent of the world's greenhouse gas emissions, so we really have no say over reducing pollution. But we do have to prepare. Just look at how much coastline we have. You were in Belize City, right?"

"Yes."

"Did you notice that it is barely a meter above sea level?" Anthony explained that the entire city sits on a peninsula made by sediment from Haulover Creek. In 1961, Hurricane Hattie leveled the city, after which they moved the capital to Belmopan. A combination of stronger storms and rising sea levels could literally swallow Belize City. "The goal of our project is to figure out how we are vulnerable to climate change, and get ready. In Caribbean nations, 40 percent of the people live within two kilometers of the coast. We need to prepare. We must not overdevelop our coasts."

"So . . . how are you going to prepare for global warming other than not building on the coast?"

"Um . . . well, that's our main policy."

"Is it working?"

"Not really. Too many people live on the coast, and more are settling there all the time. Also, we need to diversify the economy, because we're too dependent on our coral reefs for tourism. Those reefs are highly at risk—if they go, so will much of our economy."

Belize and small island nations have such a different perspective on climate change than do the United States and other larger countries. Belize's government has no discernible ability to shape global politics, and their country produces less than 0.002 percent of global pollution. They are almost powerless to affect climate change, yet they still must adapt to a warmer world. If we need to work together as a "world team," as I said in my interviews, what does that mean for Belize and small island nations?

———————

After visiting Galen University in the nearby town of San Ignacio, where I talked with a small group of students, Dennis and I charted a course directly from the capital to the Caribbean. Dennis rode a new green Trek 520 touring bike that he had pedaled a total of twenty miles before boxing it to fly to Belize. He had two rear panniers instead of my four, which meant he could match my speed with far less effort. His panniers were similar to my front pair, but his bags' straps were unfrayed, the reflective strip on the rear unripped, and, of course, they weren't painted brown to look cheap. His tires had fresh treads, whereas mine had nearly worn through, and his bike wasn't covered with faded stickers. From a distance, the two of us appeared to be on the same bicycle trip; close up, we were clearly on different journeys.

As we rode, we recounted numerous failed romances and past adventures, such as climbing in Yosemite and late nights on campus doing physics problem sets or crashing parties. Along the way, Dennis regained the playful smile I know him for. We connected my iPod to the small battery-powered speakers he'd brought, so as we biked the empty road we blasted songs by U2, Willie Nelson, Johnny Cash, Bob Dylan, and Jarabe de Palo.

As night fell we looked for a place to camp. "So, we just bike until we see a good place to stop?" Dennis asked.

"Yeah. If we can't find a place to hide our tents, we should ask at someone's house." The coastal plain had short thick brush on either side, with no places to pitch a tent. Eventually we approached a modern one-story house at the center of a flat, lush lawn. A dirt driveway lined with short palm trees led from the road to the house. Arching over the entrance to the driveway, a large sign with red-and-yellow cartoon letters read THE DUTCHMAN RANCH. On a wooden pole hung a BEWARE OF DOG sign, except the word "dog" has been covered by a plaque that instead read TIGER.

"Hey Dennis, let's ask here."

"Ahh . . . do you think he has a tiger?"

"Of course not. It's a joke."

We walked our bikes down the driveway and knocked on the door. An older man, maybe seventy years old, answered the door and leaned toward us, straining his eyes. "Hello. Dutchman here," he said.

"Hi. We're traveling by bike and looking for somewhere to camp." I gave him my business card, and explained I was on my way to Patagonia.

Dutchman, who explained he was from Michigan and now retired, welcomed us to camp on his lawn. As we walked our bikes around the house, he added, "I'll have to put my jaguar inside."

"Um . . . okay."

Dutchman opened a door to his rear porch and called, "Here, Baboon!" A cat about three times the size of a house cat and covered with black spots ran in the house, and Dutchman closed the door. "A local found this jaguar kitten a few months ago. She was orphaned and wouldn't have survived, so I'm taking care of her. She's eleven months old."

"What . . . uh . . . are you going to do when she is full grown?" I asked.

Dutchman shrugged and returned inside. In the fading light, Dennis and I extracted our tents from our panniers. As we debated tent sites, we heard a loud thud from the house. The porch door flew open as Baboon forced the latch ajar. The cat bounded toward us.

She pounced on Dennis, reaching her claws up to his chest. "AHHH!" screamed Dennis, madly pushing Baboon away. The cat then leapt on our pile of gear and ran across the lawn, dragging Dennis's helmet by its straps. About thirty feet away, she then sat down and gnawed at the helmet while holding it in her paws.

"Hey man," I said, "you're not going to find another helmet around here. Go get it."

"You go get it!"

"We'll do it together."

Baboon with Dennis's biking glove

Dennis and I ran and screamed at Baboon, who dropped the helmet, ran a circle around us, and then pounced again on our pile of gear. She next found one of Dennis's leather biking gloves, and ran off with it to another corner of the lawn.

"Dude, the biking glove's not worth it. Let's get out of here."

Dutchman, hearing the yelling, hobbled out of the rear door cursing and waving a fly swatter in his hand. Ambling over to Baboon and yelling, he raised the swatter and *thwapped* Baboon hard on her back. "Bad kitten! Bad!" The jaguar jumped and clawed at Dutchman's hands, scratching, drawing blood. Dutchman hit the cat harder.

Baboon dropped the glove and streaked off into the forest.

Dennis retrieved his glove, freshly covered with jaguar saliva, and said, "Um, I think we're going to pack up and ride on."

"Yeah," I said, "I mean, it's a great lawn, but I think we'd best be on our way."

After repacking our panniers, we biked down the road to find a small village named Maya Centre, where the locals spoke the indigenous language Mopan. A man let us camp behind a few cabins, and three of his children, all between the ages of five and ten, wrote out Mopan words for us on a sheet of paper.

Dennis looked at me and said, "I take back what I said about this being like Connecticut."

DAY 133
March 17—Placencia

The next day we rode five hours to the small town of Placencia, Belize, which lies at the end of a long sandy peninsula that runs parallel to the shore and is only a few feet above sea level. Between the peninsula and the mainland grow thick mangrove swamps. Though Placencia consists of nothing more than a single street and parallel boardwalk and has only about six hundred permanent residents, enough tourists visit to support thriving restaurants and fancy lodges. Dennis and I set our tents on the beach beneath a palm tree and paid a local woman five dollars a night for beach space, bathroom use, and a lockable shed.

Our one objective was to see the famous reefs of Belize. A number of small businesses ran dive operations, but I could convince only one to certify me to dive on such short notice. (Dennis was already certified.) The combined cost of the class (five hundred dollars) and the underwater case for my camera cost more than twice as much as I had spent the entire month of February—and I would have paid more, given what we were able to see.

My instructor, Julie, was a forty-year-old Belizean woman of African descent who usually taught guests staying at the expensive Turtle Inn, where a room for the night also cost five hundred dollars. After a day of book study and practicing with scuba gear in a pool (Dennis spent that day diving with another group), I was ready for the open water. Dennis and I joined a dozen Turtle Inn guests in a boat that Julie motored fifteen miles out to the barrier reef. On the ride we talked with those diving with us: a banker from New York; a man who was a board trustee at Yale; and a couple who owned a consulting company, plus their teenage children. Dennis, who'd had two cups of coffee before the boat ride, led the conversation in his Texan accent, drawing laughter-inducing attention to the fact that I'd biked there, unlike the rest of them, who'd flown in. The wind blew in our hair and water sprayed, cooling us despite the sun. During a lull in the conversation, Dennis leaned over and asked, "Do you ever get the feeling that we're living beyond our means?" Dennis and I were more used to hitchhiking than spending time with bankers.

"I feel like I'm on vacation," I replied, smiling. I was enjoying the typical tourist experience, but as I laughed with my fellow sailors, I also wanted to tell them what they were missing. None of them camped for the night in a town where everyone has dirt floors and few people have electricity. It both amazes and saddens me that it's so easy to travel and be a tourist without ever seeing the "other" side of the countries you visit.

———————

Sitting with my back to the water and my hands on my mask, I leaned and fell backwards, my fins briefly in the air before I dropped beneath the waves. Below me and forty feet of impossibly clear water lay the sea floor. Following Julie and Dennis I slowly descended toward it, making a conscious effort to continue breathing, in and out, never holding my breath.

On the sea floor the swell of the water slowly rocked marine life back and forth. Four-foot sponges swayed in unison like leafless trees, and long fern-like plants joined the dance. Long cylindrical sea cucumbers resisted the current, standing like sturdy tree trunks, barely moving. On the floor of this forest were the solid, rippled coral reefs.

Fish swam around us, so many that it seemed as if we were in an aquarium. A flat black fish with a yellow stripe swam by my head as a group of football-sized blue fish weaved through the forest of sponges. A dark green–and–blue fish rested near the bottom, blending in with the floor. Above, a few hundred black silhouettes moved against the rippling light blue surface. Julie, our instructor, pointed out two lobsters hiding on the floor. An eel extended itself halfway out from beneath a coral, and barracuda swam at a distance. A manta ray lay flat, its gray-speckled back perfectly matching the sand between corals. Dennis floated upside-down, a yellow fish swimming around his head.

Belize's coral reefs

I had seen such places on television, but somehow never believed they would be more impressive in person. *But there it was—a teeming ocean of life.* My excitement had gotten the best of me; I had been breathing too hard, too quickly, and was almost out of air. We returned to the surface for lunch, during which all I wanted to do was return to the water.

Coral reefs are the world's second-most biologically diverse ecosystem, trailing only tropical rainforests in diversity of life. Though they cover only 0.1 percent of the ocean's floor, they are home to one-third of all species of marine plants and animals, including at least four thousand species of fish.

The reefs themselves, which provide habitat for diverse marine life, are actually two creatures living symbiotically, polyps and algae. Polyps are tiny animals that build the structure of the reef out of hard calcium carbonate. The structure houses the algae cells, which photosynthesize, using the sun's energy to provide food for themselves and the polyps. A reef is a colony of millions of these polyps and algae living together.

Nearly every study on global warming paints a gloomy picture for the future of coral reefs. For a reason not yet understood, warm temperatures cause reefs to expel the algae that feed them, and the reefs eventually die in a process known as "bleaching," so called because the dead reefs appear pure white. In 1998, one of the warmest years on record, 16 percent of the world's reefs succumbed to this fate. Though about half of the reefs regenerated in the decade following, the rest remain lifeless. If the planet warms by many degrees, a majority of the world's reefs could die. Heat-resistant reefs might survive, but many would not, and the diversity of the oceans would plummet.

There's another threat, though, one that has to do with the way greenhouse gases affect the ocean. Carbon dioxide reacts with water to form carbonic acid; as such, the oceans have a remarkable ability to absorb this byproduct of fossil fuels. Over the past few decades, the oceans have absorbed about a third of our carbon dioxide pollution, thus keeping the levels of carbon dioxide in the atmosphere lower than they would have been otherwise. But, in the process, the oceans have become more acidic.

Reef organisms require slightly basic (the opposite of acidic) pH levels to form sturdy shells of calcium carbonate; as the oceans gain acidity, the reefs struggle to make their shells. If the oceans become overly acidic, reefs could literally dissolve. As one of my former professors had recently told me, "We're creating conditions in the ocean that we haven't seen since the dinosaurs went extinct. By mid-century, reefs might not be able to grow their shells."

In the worst—and increasingly likely—case, all coral reefs, these natural wonders of the world, would be obliterated, taking a large amount of ocean biodiversity with them.

———

Dennis and I spent another day scuba diving, capturing a few gigabytes' worth of videos and photos on my memory cards. Returning to our tents on the beach, we spent the evening at the Tipsy Tuna and the Barefoot Bar, where we drank Belize's Belikin Beer and tried to meet young single women on vacation. We weren't successful, perhaps because we spent too much of our energy trying to be funny and too little of it listening to what *they* had to say. Maybe there was a reason we were single.

Few Belizeans can afford to scuba dive and enjoy the reefs in the same way that Dennis and I did; the amount I spent for the experience was greater than most Belizeans make in a month of work. But while many in Placencia would consider it a luxury to dive in their reefs, their existence depends on them all the same, and they would still lament their loss. The allure of that indulgence keeps tourism dollars flowing into their country, providing work for all the residents—not to mention the fishing industry resulting from the reef's vibrant ecosystem. Being able to dive in the reefs may be a luxury, but their existence is not.

BELIZE
March 12–22
Presentations: 1
Flat tires: 0
Miles: 287
Trip odometer: 4,374 miles

8 HONDURAS
Storms

Honduras population: 7.9 million
Annual per capita GDP: $4,200
Annual per capita CO_2 emissions from fossil fuels: 1.1 tons

DAY 142
March 26—San Pedro Sula

Rain soaked our biking jerseys and our tires splashed water upwards, caking us with a layer of wet grime. I got two flat tires in quick succession, which I fixed under the overhangs of two-story apartment buildings along the road. Children peeked through the concrete window frames, watching me lay *del Fuego* on its side, remove the wheel and wet tire, and patch the tube. Since passing into Honduras, we had seen far more people, and more children running in the streets—a stark contrast to the emptiness of Belize.

Stressed by the rain, Dennis and I argued. We were running out of time for him to make his departing flight, and we feared the crime of Honduras's second-largest city, San Pedro Sula. Honduras, like its neighbors, had one of the highest murder rates in the world. Since this crime is concentrated in cities like San Pedro, Dennis and I pedaled nervously through the city's blocks, anxious to cross the town before dark.

When Dennis and his bike disappeared into the terminal at the tiny San Pedro Sula airport, I immediately felt more alone than I had before he joined me. He had restored the warmth and camaraderie one feels when spending time with a good friend. Biking back into town, I found myself unenthused for the road ahead, and felt a loneliness I hadn't known since I'd left Oaxaca.

Confronted with such homesickness, I followed the obvious course of action: I biked to the fire station. Of course, living up to the generosity of their more-northern brethren, the *bomberos* of San Pedro Sula invited me in for the night. And once I listed the stations where I'd already stayed, these firefighters seemed eager to outdo my previous hosts. When I told them it was my twenty-seventh birthday, they gave me a BOMBEROS DE HONDURAS T-shirt, and one firefighter showed me inappropriate videos of his girlfriend on his cell phone. In the warmth of their welcome I again forgot the solitude of travel.

In just twenty-four hours my emotions had leapt from exhaustion and frustration to loneliness and then to joy. But this was simply a faster emotional transformation than I'd already been experiencing along my journey. My days now often had a bit of each sentiment, with any current mood likely to be replaced by another as long as I kept riding—a much higher emotional volatility than I was used to. Even though I spent hours on end riding my bike, wheels turning ever forward, each day brought new surprises, with limited routine.

Biking south on the low-traffic highway, I scanned the people along the road, looking for unfriendly eyes. Just as in Chiapas, some children yelled "gringo" as I passed, but the stares I got were more curious than fearful, and my frequent check-ins with the police reassured me that the rural road would be safe enough.

DAY 146
March 30—Beyond Santa Rosa de Copán

The highway led me into rounded mountains whose flanks had been cleared for plots of agriculture and grazing. Mud buildings lined my route. On the road's shoulder sat rectangular mud bricks laid out to dry in the Honduran sunshine.

In a valley beyond the town of Santa Rosa de Copán I passed an elementary school. The students ran about in a concrete courtyard, screaming and wearing a colorful collection of dirty shirts. I noticed that all the boys had short hair, and all the girls had shoulder-length hair, as if everyone visited the same barber. The single-story school building, with turquoise-painted walls and a metal corrugated roof, was shaped like a big U, with the courtyard facing the road. After a group of students noticed me watching them, a small crowd pushed against the fence separating us.

A middle-aged teacher waved me in and, once my business card was in hand, asked me to give a short talk to the students. "You know about global warming?" I asked him.

"Yes, of course, the heating of the land, the ozone layer and all. It's a big problem."

Such was the now-common reply. Perhaps half of the Latin Americans I mentioned global warming to immediately cited the ozone layer.

"Gather round!" he yelled to the students. Soon a hundred dark-haired kids, probably between seven and nine years old, stood and stared at me.

Nearly every school I had visited so far had been a private school, with students who didn't live in poverty. I hadn't yet talked to a truly poor school about climate change, and these kids lived in one of the poorest countries in the Western Hemisphere. *What should I say to them?* The truth is I was embarrassed to talk about global warming to the poorest of the poor—those who use almost no fossil fuels, and who want to use more.

What sort of environmental message would be worth sharing? Should I tell them not to throw trash in the street? Yes, garbage was a problem, but those students were probably excited any time they could afford a soda bottle. I had noticed there was less trash in Honduras than in Mexico, mostly because Hondurans can less afford drinks packaged in plastic or glass bottles. I had actually seen many beverages sold in plastic bags, as thin plastic is cheaper than bottles—and yes, those bags were scattered about the countryside. *Should I tell them not to cut down the forest?* Deforestation is probably the biggest environmental problem in Honduras, but many people survive on agriculture; they clear the forests to grow food.

"Who here has a bicycle?" I yelled so the children could hear.

Four students raised their hands.

"I am biking from California to the tip of South America. Where is California?" About twenty students blurted out various answers; the other eighty just stared at me. I told them California was north, five months by bicycle, and then segued into the best talk I could think of on the spot.

"I am biking because the bicycle is a great form of transportation. But I am also traveling around to talk about a problem we all need to work together as a world to solve.

"Each of you shares this courtyard with the other students in this school. And because you share this courtyard, you shouldn't throw trash in it; it's not just yours to trash. In the same way, we share the atmosphere with everyone on the planet, and we shouldn't pollute it, as that would be the same as throwing our trash into it." I saw that, though some students continued to stare at me, the attention of many had wandered. A boy played with a girl's hair, and two girls started talking to each another.

"And it's very important to study hard in school. If you do well in school you can do anything." I wondered if that was true—I doubted good grades were a guarantee of a job in Honduras. They could make far more money by traveling to the United States and working as a janitor than they could with a high school diploma in Honduras.

Students at El Progreso, a school near Santa Rosa de Copán

Sensing that I was losing their attention, I directed the students into a large group so I could take a picture, then thanked the director and pedaled away.

As dusk approached, I looked for a family willing to offer me a campsite for the night. Unlike in suburbs in the United States, many people were out, either in their front yards or walking the streets, so I had many candidates to approach. I found I had the most success asking a man for assistance when he was with another family member, as he would do his best to show off his generosity.

My luck for that evening came from Melvin, a man in his early thirties who held his daughter's hand as they enjoyed a walk together at the end of the day. He led me to their small mud-and-brick house, a structure about twenty-five-feet long and fifteen-feet wide. The roof had curved tiles instead of corrugated tin metal or palm leaves. Their floor was dirt, and they had no electricity.

I usually had ingredients for dinner with me before I sought out my next campsite—I never expected my hosts to feed me—but I had none that day. So I was thankful when Melvin's wife, Rosa, invited me to join them for dinner: tortillas with beans.

"Where do you come from?" Rosa asked me as she cooked a pan of soupy beans on a small wood stove.

"I biked here from California."

"How many days?"

"Days? About five months!"

"Mmm . . . why?"

"I like to travel by bicycle. Also, I'm riding to raise awareness of global warming." I gave her my business card.

"Oh I see, they pay you for this."

"No."

Rosa contemplated this thought, and appeared to not understand that I was doing this project without getting paid. She asked no more questions and continued cooking over the small fire in the corner. The tortillas she eventually served were smaller, thicker, and less tasty than the tortillas I'd had in Mexico. The beans were a runny mush, and the portions small. I asked if they eat meat often; she replied they occasionally eat the chickens they own, but they eat mostly beans and rice because meat is more expensive. I guess that wasn't too surprising—the average citizen of the world eats less than half as much meat as the average U.S. citizen does, and the poorest of the poor eat far less. The raising of livestock, because of the agriculture and deforestation that is required to support it, accounts for a large percentage of greenhouse gas pollution. Melvin and Rosa have a small carbon footprint for many reasons other than just their lack of electricity.

Rosa offered me water. I usually didn't drink untreated water—my stomach was not yet as hardy as the locals'—but I was out of both water and iodine purification tablets. Rosa told me the drop of lemon juice she'd added made it safe. I doubted that was true, but I was thirsty, so I accepted her lemony water.

Though darkness enveloped the house, they lit no candles. Could they not afford them? The little light they had came from the flickering fire of the stove. Melvin and Rosa spoke little. Their daughter, maybe three years old, said almost nothing. When I asked Melvin if he had any books, he pulled down from a high shelf their only book, the family bible. It looked dusty and infrequently read. I felt almost a mental suffocation in the house—the room was dark, few words were spoken, and there was almost nothing to read and no games to play. I wondered how their daughter could learn in such a limited environment, at least limited compared to the wealth of information and stimulation I had growing up.

To my surprise, Melvin turned on a television that sat on a small table next to a wooden couch. The TV was a small black-and-white box, powered by a car battery. He explained to me that he paid fifteen lempira, about one dollar, to charge the battery once a week. We watched *Betty La Fea (Ugly Betty)*.

Melvin then asked something I would hear more than once during my trip: "You have nice homes in the U.S., don't you?"

"Yes, we do. How do you know?"

"I see them on television."

Wow—everyone has a portal into the modern world.

Melvin explained he had three cousins in the U.S., and that it was nearly impossible to get a job here. He then told me of the few jobs he's had in the past decade. The best paying, working as a security guard, was also the least satisfying. "I liked building a road the most," he said, describing the road he built with others in his town. "But when the road was finished, I had no more work." He now survived on growing corn and beans behind the house.

Melvin, holding a machete, and his television powered by a car battery

After offering me a spot on the floor, Melvin went to bed. I laid my small inflatable Therm-a-Rest mattress on his dirt floor by the television.

In the morning Melvin gave me a tour of his land, where he grew corn and beans on less than one acre. I asked him many questions: when he harvested his corn, how much he harvested, where he stored it, how he sold it, how many chickens he had, where his other family members lived. He answered all questions but asked few in return, other than the price of *del Fuego*.

"Is the climate here different in recent years than in past decades?"

"The rainy season and dry season used to be very divided," he said. "It would rain during the rainy season, and be dry during the dry season. Now we can have rain or drought at any time of year." I had heard a few other Central Americans make the same claim, that the weather is now more variable.

"Why do you think that is?"

"People—we're changing something," he said, looking up.

"Do you think we can do something about this change?"

"No," he replied, echoing a fatalistic view I often heard. Many people I talked to in Latin America seemed to agree that humans are changing the environment and climate, but that there is little we can do to reduce our impact.

"What happened when Hurricane Mitch came through here?"

"El Mitch?" he said, and his eyes rose to meet mine. "We lost all of our crops, and we went hungry."

———

In 1998, Hurricane Mitch struck Central America, taking more than ten thousand lives. Most died from drowning—mud houses don't survive well in floods, and many people's houses were swept away. Unlike in the developed world (with the notable exception of Hurricane Katrina), where hurricane damage is usually measured in dollars, in the developing world hurricanes are measured by the number of lives lost.

Climate models suggest hurricanes will become more powerful in a warmer world. But even if *hurricanes* per se don't become more powerful, storms in general are definitely expected to be more destructive, mostly because of heavier rainfall. The reason is that warmer air can hold more water: for every degree Fahrenheit warmer air becomes, it can retain about 4 percent more water. The result is more rain—balanced, of course, by more evaporation. But this increased rain doesn't necessarily mean just *more* storms of the same intensity as before; both climate models and scientific observation suggest that the increased rainfall will instead be concentrated in the same number of *heavier* storms. As the Earth has warmed over the past fifty years, in some places storm sizes have increased even though total rainfall decreased—meaning that the rain fell in fewer, heavier storms. For instance, in the United States, where we have the best weather data, the amount of rain falling in the heaviest storms (the storms in the top 1 percent by total rainfall) has increased by 20 percent in the past half century. The result of such weather is both more flooding and more droughts, which could mean more years of failed crops for subsistence farmers like Melvin and Rosa.

Some economists argue that we should focus on helping the poor develop economically, so that when a warmer world comes, people can afford to adapt. This idea makes sense: if Melvin and Rosa's children had more wealth, then they could build a stronger house that could better survive storms. Perhaps they would be able to buy food imported from Canada, or maybe they would invest in genetic engineering to create crops that can survive a warmer Honduras.

But to fuel such economic growth with our current technology would mean burning more coal and gasoline, which would just exacerbate global warming. More than two billion people—people like Melvin and Rosa—currently live on less than two dollars a day and use almost no fossil fuels. Before folks like them escape poverty using the same technologies that my country used to develop our economy, we will likely pass the point of no return: when the world is on track to warm by more than what most experts consider "safe" (about 2 degrees Celsius, or 3.5 degrees Fahrenheit). Without affordable, abundant, carbon-free technologies, there is simply not enough atmosphere left for the Melvins and Rosas of the world.

When I rode away from Melvin's mud-and-brick home the next morning, thinking about climate change and its implications for families like his made me feel both guilty and sad. I felt guilty because humans have already emitted perhaps half of the carbon dioxide the atmosphere can absorb if we are to keep the Earth's warming within a safe level, less than 2 degrees Celsius—and the majority of that pollution comes from rich countries like mine, which represent less than one-fifth of the world's population.

Before I left I gave Melvin my headlamp, partly in thanks, partly because it pained me to think of them in the dark every night. I wish I could have given them an electric light, and connected their house to the electric grid. I couldn't do either, but I feel we need to somehow help Melvin and families like his—the rest of the world's poor—improve their lives without overtaxing the atmosphere in the process. Melvin and Rosa deserve a more stable, comfortable lifestyle, which means using more energy, not less, and I am morally opposed to preventing them from getting it. If we have to choose between denying them energy or letting climate change occur, I would vote to let climate change continue unabated. I hope, though, that we don't have to make that choice.

DAY 157
April 10—Tegus

After more riding through the mountains of Honduras, I arrived in the Honduran capital of Tegucigalpa, or "Tegus," as the locals call it, where I stayed with Martha, the mother of a friend from college, and her husband, Carleton. Both were Hondurans who had lived in the United States for many years before returning to their native country. The two lived in a house at least five times the size of Melvin's home, on a hillside overlooking the city. They had modern indoor bathrooms, high-speed Internet, potted plants, and three well-fed guard dogs who barked loudly at intruders. (One of the dogs had recently put a previous guest in the hospital. I tried to be on good behavior.) Thick metal bars lined the windows and a security wall surrounded the property—as if the house were a prison.

Carleton, a clean-shaven, short-haired man in his fifties, offered to take me on a tour of the city. He spoke perfect English, articulating his sentences clearly, and had a mechanical but friendly smile. He was the principal of a successful private school (that happened to be closed for vacation).

We drove in his pickup truck down the hillside toward downtown Tegus. The city of about a million inhabitants filled a large bowl-shaped valley. Shacks lined most of the city's fringes, but a few trees still grew among them. The center of town, which I had biked into the day before, lay at the bottom of the bowl, along the banks of a small river. The main plaza felt small and crowded. I noticed the cathedral could have used another layer of paint. Many of the stores had armed guards—even a donut shop had a man in uniform with a gun—and on the street I felt the same eyes I saw in Chiapas, the eyes of crime and fear. I was glad to be staying in Carleton and Martha's well-protected house.

Carleton drove me down to the river, which ran through the downtown in the bottom of the valley. I asked about Hurricane Mitch. "Well, when the hurricane came through, two million Hondurans—a third of the country—were temporarily homeless. Sixty percent of the bridges in our country were destroyed. When you cross a bridge now, you can usually see what aid agency helped build it." He slowed down near a two-hundred-foot-long bridge crossing the Choluteca River. "The government of Norway helped finance this bridge," he said, pointing to a plaque on the right side of the bridge. "You see the building over there across the river? When Hurricane Mitch came through, water reached its third floor. Beyond that building you can see the bare hillside. That used to be covered with houses. They were all washed away." The majority of the damage was not from the hurricane's winds, but rather from the heavy rain that accompanied the storm.

"You might argue that those slums shouldn't be built on the hillside in the first place," I said, pointing to where the houses had stood.

"Yes, but you could say that about those buildings too"—he pointed to the edges of the valley, where shacks lined the hillsides. I almost wanted to add that you could say that about the entire city.

I asked Carleton about the bumper sticker on his car: CARLETON ALCALDE. Carleton replied he had run for mayor—*alcalde* means "mayor"—explaining that he'd run on an anti-corruption platform, but lost to a man "who received illegal funds from the government." He said that when they went on television for debates, he didn't get enough time to talk because he hadn't paid the television company enough. "But," he said with a smile, "I bet I could get you on television without having to pay anyone."

———

The next morning, thanks to one of Carleton's friends, I had a twenty-minute live interview on *Buenos Días Honduras,* the Honduran equivalent of *Good Morning America.* As I spoke into the microphone in a set that mimicked a modern living room, I wondered how many Hondurans like Melvin watched my image on a television powered on a car battery, or in a house with dirt floors, and what they thought about the gringo biking and talking about global warming. Did they fear the stronger storms of a future world, or were they transfixed by the idea of someone being so wealthy he could bike across continents for no pay? Maybe they watched television to escape their current life and dream of a richer one, and maybe I—a gringo on a long vacation standing in a simulated modern living room—provided an image of that richer life.

I can't recall when exactly, but at some point in my journey through Central America, the poverty got to me. So many people in Honduras live similarly to Melvin and Rosa, in homes with crude plumbing, and sometimes no lights, candles, or even a bookshelf. Their food is simple, and employment options are almost zero. Such a life felt incredibly limiting. But in one corner of many homes is something that made me sadder and yet hopeful too: a television, a tiny window into a life with more opportunity.

Carleton Corrales

An interview on *Buenos Días Honduras (Good Morning Honduras)*

HONDURAS
March 23–April 12
Presentations: 1
Flat tires: 5
Miles: 509
Trip odometer: 4,883 miles

9 EL SALVADOR 🚲
The Scars of War

El Salvador population: 6.3 million
Annual per capita GDP: $7,100
Annual per capita CO_2 emissions from fossil fuels: 1 ton

DAY 148
April 1—Guarjila

After officially crossing the border into El Salvador—which meant waiting in line for half an hour to get my passport stamped and exchanging my lempiras for the U.S. dollars that El Salvador uses—I started riding toward the small town of Guarjila. A few different people had emailed me saying I had to go to this town and meet John Guiliano, who "has biked everywhere," and "knows everyone."

In the emails I learned that John, originally from New York City, had spent the past two decades in El Salvador. He had lived there during the civil war, which he spent in the mountains with the guerilla army. He also has biked across the United States and Central America numerous times. In his invitation to visit him he wrote, "Just ride into Guarjila and ask where John Guiliano lives."

In addition to hearing about John's extensive biking experiences, I also wanted to hear firsthand accounts of El Salvador's war. I had read studies suggesting that in poorer nations droughts make civil wars more likely, meaning there may be a link between conflict and climate change.

After a night at the fire station in the city of Chalatenango, a few miles outside of Guarjila, I biked up a one-lane paved road lined by short trees sparse enough to offer a view across the landscape. El Salvador is on the drier, Pacific side of Central America, and the terrain was strikingly different than in Honduras. The brown grass all across the mountains awaited the impending rainy season, and the sparse trees were shorter and had fewer leaves than those I'd seen in Honduras. At the top of a hill a number of buildings, some made of wood but most of concrete, surrounded a grid of both dirt roads and paved streets. When I asked a woman where John Guiliano was, she pointed me to a store. Entering, I saw a group of U.S.-looking high school students eating breakfast.

"You," said a bald white man with a strong frame, probably in his forties, speaking with a New York accent. "You need to eat. And eat as much as you want." He spoke forcefully and almost aggressively, as if I had no choice in the matter. He introduced himself as John, and said he had to go and "get Delmy ready for the presentation." I sat down with a few of the students, who told me they were visiting from a private Catholic school in Indiana and were learning about El Salvador on their spring break. I ate some scrambled eggs and toast, then followed them a few blocks to a field, where John sat on a bench next to an attractive Salvadorian woman in her mid-thirties. We sat on the grass in a semicircle. A clearing in the trees provided a view to the mountains rolling toward the Honduran border, and loud insects buzzing in the grass forced me to lean forward to hear.

John started speaking slowly and emphatically. "So, we're going to have Delmy tell her story, and in the process begin the story of El Salvador. So I introduce to you Delmy, Dr. Delmy."

"Good morning," she said in English, and then continued in Spanish. "It's difficult to tell my story, because I have to remember the war years." She paused, and John translated her words to English. "My name is María. *Delmy* is the name they gave me during the war."

"We fled to the mountains. We had no house, no tent," Delmy said, pausing after each sentence for John to echo it in English. "My older brothers joined the guerilla army, my parents worked as cooks. My oldest sister was working as a nurse, and I helped her. I was just a teenager. I had no medical training. I hadn't even finished elementary school. When I was sixteen, I had to amputate a man's hand. The head nurse was busy, and I had to do it myself. I worked hard to keep his fingers, and managed to keep his thumb . . ." She spoke this last sentence with a bit of a smile. John interpreted Delmy's words with a slow contemplative weight, as if they were his own.

"We lived in the mountains, didn't know what would happen day to day. Many times we didn't have anything to eat. Two of my brothers were killed. I had a son, and his father was killed two days after my son was born." She paused and curled her lips. The students from Indiana continued to sit in attentive silence. "It was a hard time, but also we had so much solidarity. We were all young people like you. We worked together, and we gave each other affection because we had nothing else to give. In the night we would dance together. The war was very difficult, but in some ways beautiful. We had nothing." She spoke her words with a mixture of nostalgia and regret.

El Salvador's civil war pitted indigenous landless peasants against the landholding elite, pitting people who claimed to be Marxists against the government. Similar conflicts have played out all over Latin America. The Spanish are partly to blame, as their colonial leaders established a semi-feudal system in which a small landholding elite ruled over lower classes. Many attempts to overturn this inequity have led to bloodshed and civil war. On the one side are peasants who demand more rights and a redistribution of wealth. On the other are people who fight for what they say is order and stability. Even in countries not at war, a high crime rate suggests that the struggle continues, and barred windows, wealthy homes with multiple locks, and security guards demonstrate that tension between the classes continues as well.

El Salvador's war began in 1979. But what had started as guerrilla fighting escalated into full-scale warfare following the 1980 assassination of Archbishop Oscar Romero by right-wing death squads. Romero, along with a few other members of the Catholic Church, had advocated for more rights for the landless people of El Salvador, and embraced nonviolent opposition. Ironically, his death spurred armed resistance against the government, and the government responded with even more violence.

The war raged on for over a decade, claiming perhaps seventy thousand lives. As foreign governments, including ours, saw the conflict as a Cold War struggle of capitalism versus socialism, the Carter, Reagan, and Bush administrations provided millions of dollars of aid to the Salvadorian government, which greatly prolonged the fighting. The poor, such as Delmy, were the worst affected—in a few cases, government militias massacred entire mountain villages. After nearly thirteen years of strife the war ended in 1992 when the two sides signed a peace agreement borne of extensive negotiations. Although there are still deep divisions in the country, today representatives from the former warring parties now peacefully run for elected office.

After the war, Delmy decided to go into medicine, but before she could go to medical school she first had to finish high school, then college. She described what it was like to learn basic reading and science as an adult. She is now a certified doctor, and runs a clinic for the town of Guarjila, which was rebuilt after the war. She spoke her final words with a quiet pride. Her story finished better than that of many Salvadorians.

I spent two days in the town, gradually hearing more stories from John about El Salvador. John had come to the country in the early 1980s as a volunteer for the Catholic Church, providing humanitarian aid to the resistance. At some point—though he didn't say when—he found himself fighting in Delmy's army, and he spent years in the mountains. John now runs a youth group for teenagers in Guarjila and also hosts students from the U.S. to learn about El Salvador.

Speaking with his "tough-guy" confident New York accent, John explained to me why the war ended. "Money is what drives wars. The reason the war ended is because both sides ran out of money and people." John then added scathing remarks about the leaders of the war, both of his army and the government's.

After the war, and before starting the youth group, John embarked on a ten-thousand-mile speaking tour by bike, riding from New York to El Salvador and giving talks about El Salvador's conflict. This commonality we shared was a large one, and I appreciated being treated like a brother by this man, who saw to it that I was well fed and comfortable during my time in town. But it also felt odd to share in one experience, bicycle travel, while having no commonality regarding living through an armed conflict, something I hope I will never come close to experiencing. All the same, we connected well. I also noticed that, when he looked at me, I saw a certain longing for the open road.

DAY 150
April 3—San Salvador

John fulfilled that longing for a time by joining me when I departed Guarjila. "People used to ride with me out of the cities I visited . . . I just have to join you," he said, mounting his racing bike to pedal the thirty miles to San Salvador, the country's capital. We followed a highway with fresh pavement and a wide shoulder, allowing us to bike side by side. As we rode, John shared more stories of bicycle travel: he'd carried a machete to clear forest so he could camp; when he pedaled into Mexico City to give a talk he received a police escort (though he didn't tell me why); he met his wife while biking across Mexico, and she liked him even though he had grown a dirty beard and lice crawled in his hair; and he'd met many other women who were also unfazed by the beard.

"You know," he said, "when you travel by bike, everyone's willing to help you. I once crossed the U.S. on no money at all. It's like a gift." He paused. "It's also like you're going into debt to the world."

"Well, you're certainly trying to pay it back! I fear that if I ever own a house, I'll have to treat every future guest like a king."

"But you know what? It's the greatest debt to repay."

We passed a roadside sign reading in red, white, and blue, Tu Dinero al Instante . . . Gigante Express . . . Remesas Familiares, with a picture of a wad of American dollars, advertising a service to send money from the U.S. to El Salvador.

"The entire economy here is built on *remesas*," said John scornfully. *Remesas,* or remittances, is the money workers in the U.S. send home to their families in Latin America. He explained that such "income" is now over 15 percent of the country's GDP. "It's just a *remesa*-and-spend economy. We now have some of the highest-rated shopping malls in the world. Everyone just waits for their check and then spends it."

His words resonated with my brief experience. When I crossed the border from Honduras just a few days before, I'd immediately noticed that El Salvador was wealthier—buildings were newer, the roads were freshly paved, the Internet café had newer computers, and I could buy chocolate milk at the convenience store. Although the country is poor and has a violent past, the economy is growing, and one source of growth is Salvadorians working in the United States. Though John may be critical of the consumerism, and he may be right that it isn't necessarily the ticket to happiness or a stable economy, Salvadorians are probably quite pleased to be a bit wealthier.

—————

After John and I parted and he headed back, I spent a day in San Salvador, a busy, dirty metropolis where far too many stores had armed guards. As I'd seen in other towns in El Salvador, I noticed numerous middle-aged people with missing limbs or who walked with a limp—visual reminders of the scars of war.

I gave a presentation at a wealthy private school and stayed with a bike mechanic friend of John's named Marvin. He didn't just put me up: he also spent half a day disassembling *del Fuego,* cleaning out its hubs and removing the dirt that had caked itself on every moving part. As the ball bearings within the wheels had become dirty, he cleaned them and added fresh grease. To my question of whether the bike would make it all the way to Argentina, he replied, "probably," but shrugged his shoulders. I offered to pay. "No," he said, "John has already paid."

DAY 152
April 5—Perquín

Setting out on a smoother-rolling *del Fuego,* I left the capital to bike across the rest of El Salvador, a country no larger than Massachusetts. Cone-shaped volcanoes dotted the landscape, and fields of farmland climbed up the mountainsides, making the mountains appear as if they wore dresses sewn together from many patches. The only trees stood near the mountaintops, where the terrain was too steep for agriculture. El Salvador's forests have been mostly cut down, and only 14 percent of the country's land is still forested. John had explained that war had brought environmental destruction, as people had burned and cleared land without considering the future.

While I'm sure the war played a role, the bigger culprits are probably poverty and overpopulation, as the deforestation statistics for El Salvador equate with those of the rest of Central America. Twenty percent of Central America's forests had been cut down in the fifteen years before my journey; only sparsely populated Belize has avoided major deforestation. And deforestation does more than just threaten biodiversity; it also accounts for over half of the region's greenhouse gas pollution.

I spent my last night in El Salvador in Perquín, a mountain town with cooler air situated in the still-forested hills. The town's museum memorialized the war, though it told the story mostly from the perspective of the guerilla fighters. My museum guide, José, walked with a limp he said was from the war.

"This wall shows pictures of people who lost their lives." We paused in front of images of young people, old people, influential people. The photographs were now twenty years old, their edges curled, the images fading yellow. Another room was full of weapons—rifles and artillery, all rusting. An exhibit showed propaganda posters from around the world supporting the resistance. "There is nothing noble about war," my guide said as he limped, his left leg not strong enough for a full stride. "No glory." He then showed me a crater left by a five-hundred-pound bomb.

I felt so lucky to not live in a country during a civil war, to not lose a limb or walk with a limp for the rest of my life, to not lose friends when an enormous bomb explodes in my town. Through much of the world's history, a man in his twenties such as I would be expected to join the army—to fight, and maybe die. Perhaps I'm optimistic, but the world certainly seems safer than it did just a few decades ago, and much safer than in centuries past. I hope the major wars that shake the world are behind us.

Unfortunately, global warming does pose a security risk, and has caught the attention of many national security advisors. The principle reason is drought, as droughts have been linked to civil wars in poorer nations. Now, this isn't to say that drought causes conflict, but rather that droughts can increase stress, worsening situations. One study surveying Africa came to a depressing conclusion: in the 1980s and 1990s, twenty-nine sub-Saharan countries had civil wars. The study claimed the best predictor of whether a country would go to war was drought; other indicators, such as ethnic diversity or the structure of the government, were less important.

Indeed, the conflict in Darfur arguably may have been influenced by lack of rain. Before 1980, herdsmen and farmers in the region had coexisted peacefully. But, as rainfall decreased through the 1980s and 1990s, farmers started fencing off their plots of land to keep out grazing animals, which created tensions with the herdsman, who struggled to find feed for their herds. Conflicts between the two groups escalated to the point that herdsmen, armed as the Janjaweed and backed by the government, began killing farmers.

Unfortunately, the regions of the world most likely to suffer future droughts—parts of Africa, Central America, and Mexico—are also poor, and have large populations of subsistence farmers. I think of the machete-armed men who robbed Gregg and Brooks in the poor Mexican state of Chiapas. Chiapas had a violent peasant uprising a decade before I biked through it. If people in Chiapas can no longer grow food, will the state again cascade into chaos?

Of course, we shouldn't overly emphasize climate. That droughts contribute to war is true only in poor locations. If we solve poverty, droughts will be less of a problem. Focusing on climate change instead of poverty, then, seems like perhaps the wrong focus.

Another security concern that could result from global warming is the increased potential for mass migration. If land is no longer arable, people are likely to emigrate in huge numbers. Northern Africa's residents may clamor to be accepted by the European Union, and Mexicans and Central Americans may flood the United States in numbers far greater than they already do. While it's possible such immigrants could be peacefully absorbed, another result might be unrest, increased crime, and even terrorism. Plus, migration might also be sparked by sea level rise. In Bangladesh, an extremely poor nation, tens of millions of people may have to evacuate their homes if sea level rises only a few feet, which it is projected to do this century. But these residents may have nowhere to go; India is already building a six-foot concrete wall along the border with Bangladesh to keep out immigrants.

───────

Leaving the Perquín museum, I followed a dirt road into a forest of short evergreen trees to look for a campsite. Fifteen years earlier, Marxist guerrillas had hidden in these trees; such fighters and their families were now the residents of the nearby town. So it was a surprise to encounter on this dirt road four figures emerging from the woods, crackling the dry grass beneath the trees.

At first I was afraid, but then I recognized Josue, an eighteen-year-old boy I had met earlier at the town's Internet café. Josue called out, "¡Hola, David!" and walked up to me with smiling enthusiasm while his friends followed. They wore school uniforms—khaki pants and white-collared shirts—but now their collars were up and the shirts untucked. When he yelled out, "Take a photo!" his friends threw their hands in the air and echoed his friendly demand. I slowly noticed that Josue's right hand held a mostly empty bottle of vodka. I laughed out loud: these kids were *smashed*. I then saw that one of Josue's friend's had a small portable DVD player.

Still laughing, I pointed to the DVD player and asked, "*¿Remesa?*"

The boy said in Spanish, "Yes, from remittance. My mother is in South Carolina." Then, in broken English, "She wants better life for me!"

Josue and his friends offered me the last of their vodka, which I gladly accepted, and they returned to town, leaving me to camp by myself in the forest. Half an hour later, lying with a slight buzz in my sleeping bag, I stared at the ceiling of my tent with confusion. My experience in El Salvador had shaken me, and the stories of war and its visible scars had left a deep impression. But my last memory of El Salvador was these kids laughing and enjoying themselves in the forest. What is the future of this country? Will coming decades bring growing *remesa* checks and better DVD players? Or will these forests again be populated with chaos, providing more stories like Delmy's?

Droughts encourage civil war only in societies dominated by subsistence farmers, and I would guess that El Salvador, though still very poor by U.S. standards, is now wealthy enough that the country probably won't descend into violence. Of course, I may be overly optimistic. A severe drought could still create major unrest in El Salvador, promoting, at the very least, more migration to the United States. It could be ugly.

Nonetheless, even though I am very clear about the ways climate change poses significant security risks, I again felt confused about my mission. Climate change is not as bad as some of the other challenges facing humanity. If I could do just one thing: end all future wars, or prevent climate change, I would choose to end war. If my choices were to eradicate poverty or prevent climate change, I'd end poverty. Obviously, climate change may exacerbate some forms of poverty or increase the likelihood of some wars. It also might be possible to reduce poverty and the chances of war without fully addressing climate change. Maybe that was the confusion I felt but couldn't yet articulate, lying in El Salvador, staring at the ceiling of my tent. Why focus on climate change if there are bigger threats to humanity? Why focus on climate change if there are bigger drivers of poverty and conflict?

The problem is the immediacy of climate change as compared to these other issues. Climate change is, by its nature, slow moving. The global atmosphere takes decades to respond to higher carbon dioxide levels, and it takes humanity decades to change the way we use energy. These two inertias—of the atmosphere and of human society—mean that if we want to avoid disastrous climate change decades in the future, we must act now. If we want a peaceful and prosperous world to leave to future generations, we must address issues of poverty and war at the same time that we cut carbon emissions. The question is, are we going to do it?

EL SALVADOR
April 1–7
Presentations: 2
Flat tires: 0
Miles: 179
Trip odometer: 5,062 miles

10 NICARAGUA
The Scientist

Nicaragua population: 6 million
Annual per capita GDP: $4,000
Annual per capita CO_2 emissions from fossil fuels: 0.8 ton

DAY 160
April 13—Nicaraguan Countryside

Five men accosted me at the border, each with a calculator and a handful of cash, eager to exchange my bills for Nicaraguan córdobas. I bargained one off the other, not so much to get a better deal as to make sure I wasn't getting taken advantage of.

The similarities between the two Central American countries I was straddling felt disorienting—crossing the border into Nicaragua felt like returning to a room I knew well and finding the furniture rearranged. Nicaragua had also been torn by conflict in the 1980s, but here the socialists, known as the Sandinistas, actually took power. In response the United States, worried that it faced another Cuba, funded the Contras, an armed organization that fought the Sandinistas. (The Contras were partially funded by selling arms to Iran, hence the "Iran-Contra" scandal.)

But I was also disoriented by the fact that here I crossed borders so often. I almost wished that the Federal Republic of Central America—a country that existed for a decade after the region's independence from Spain in the early 1800s—had never splintered into these various small republics.

It was like traveling from New York to Georgia if each state were a different country. Bicycle travel was almost too fast to cross the region—just as I got used to the currency and the local customs of one country, I'd arrive at the border to the next. The experience was a bit jarring, and tiring; I found myself looking forward to the more reasonably sized countries of South America.

———

The afternoon after crossing the border, as I biked through a savannah-like landscape, an attractive woman named Cristina invited me to join her and her brothers at their town's pool. After paying twenty-five cents to enter the enclosure I at first balked at the thick brownish blue water, but then I jumped in with the rest of them, letting the cool dirty pool refresh me from the day's heat. Speakers played lively music from the 1980s—"Girls Just Want to Have Fun," "Ghostbusters," "Eye of the Tiger." I chatted with Cristina, who gave me a big smile and had lively eyes. She was about twenty-five years old and swam in a green tank top that said SEXY 87, as if she were player number 87 on team Sexy. As I had first noticed in Mexico, many of the clothes worn throughout the region appeared to be discarded garments from the Salvation Army, clothes once intended for English-speaking markets. Cristina explained that she was studying law and wanted to go to the United States. "Can you help me go?" she asked.

"No," I said flatly, explaining that I wouldn't be in my country for another year, and that I wasn't sure how she thought I could assist her. But I was excited to talk to her—it was rare to meet a woman my age who wasn't married. I had gotten into the habit of asking a woman I'd just met if she had children, and how many, as such seemed like the socially acceptable next question to ask in the conversation. It would be awkward and probably improper to ask the same question of a woman I'd just met back home, since in the States we don't have the same expectations of starting a family so young.

As Cristina realized I was not a ticket to anywhere, she started to lose interest in me. Her friendly green eyes dropped their gaze. I found myself wishing to say, "Yes!" to regain her attention, but I didn't.

After swimming I was invited by Cristina's brothers to visit their family's land, and I rode the quarter mile to four tiny brick-and-mud houses with dirt floors. "What happened when Hurricane Mitch came through here?" I asked José, Cristina's brother.

"Thankfully no one died, but we had to rebuild these houses," he said, pointing to the adobe structures.

———————

I left my swimming friends and continued on. Farther down the road, I encountered two men riding steel bikes, of the style I had seen ever since crossing into Latin America. The horizontal tube between the handlebars and the seat was reinforced with an extra bar, perhaps to provide extra strength for the second passenger who often sat on the tube. "Where are you going?" I asked.

"To go swimming. There's a stream up here. Are you from the United States?"

"Yes."

"I'm going to go there next month!"

"Really, how?"

"A *coyote*."

"How much will that cost?"

"About five thousand dollars," he said matter-of-factly. The men then turned off the road for a swim, on bikes that probably cost eighty dollars. I wanted to ask how he got the five thousand for the *coyote;* most likely it had come from a family member in the United States.

Nicaragua is one of the poorest countries I would visit, on par with Honduras, Guatemala, and Bolivia. Yet as I rode, the people I met somehow made me optimistic about the country. Perhaps I felt this way because Nicaragua was still rebounding after many years of conflict, and the economy had grown since switching from a Soviet-style economy. Or perhaps it was because the roads were newly paved, and I had a smooth shoulder to ride on. Or maybe because I'd entered Nicaragua during Semana Santa, the week before Easter, when all of Latin America was on holiday. Everyone seemed to be on the way to a swimming hole, a river, or a pool.

DAY 162
April 15—Managua

The Managua airport was a single building, with one terminal built along the main highway and a parking lot smaller than those found in front of most U.S. supermarkets. I was there to meet my father, who'd join me for the next two weeks to bike from Managua, Nicaragua, to San José, Costa Rica.

My dad walked out of the baggage claim area rolling a suitcase behind him, carrying a large maroon duffel bag, and wearing a blue bike jersey beneath an unbuttoned long-sleeve collared shirt. His height is the same as mine, and we have the same build—it's impossible to walk next to him without people knowing instantly he's my father. After hugging him in greeting I said, "Dad, you don't have to wear your helmet yet."

"Well, it was the easiest way to carry it from the plane," he said. "And yes, I'm already doing my best to embarrass you."

My dad and I have had, in many ways, very different lives. His father, his family's sole breadwinner, worked at a gas station his entire life and spent large portions of his income prospecting for oil. (They lived in western Michigan, which isn't known for its oil deposits.) My dad picked strawberries or other crops in the summers during middle school and high school, and then put himself through college—at the school closest to home—working various jobs while double-majoring in chemistry and biology. By comparison, I spent my summer at various sports camps or music camps, and had all the Transformers a young boy could want. Having grown up with different expectations and more affluence, I was able to attend a fancy private university across the country from my home. When he finished college, my dad pursued a PhD in zoology; eventually he studied and taught the science of how birds learn to sing and became a professor at the University of Massachusetts. When I finished college, I pursued a master's degree and decided I wanted to bike to the far end of the world.

Of course, it's fair to blame my dad for some of my aspirations. He enabled this trip. Around the time he retired from the University of Massachusetts, at fifty-seven years old, he decided he wanted to bike across the United States, but not alone. So he bought me *del Fuego* on the condition that I would ride it across the country with him. For two and a half months we biked from Virginia to Oregon, riding forty-five hundred miles on a route that spanned ten states. The goal of our adventure was not only to bike, but also to listen to birds. My father has dedicated his life to the science of birdsong, so every morning of our trip we would wake up early to listen to "the dawn chorus." Or, rather, my father would wake up early, and I would continue sleeping. "Four in the morning is when birds sing the most interesting songs!" he'd say with enthusiasm. "It's also the best time for sleeping," I'd reply.

Growing up with my professor father, a world expert in bird-song, gave me a window into the world of science. To be successful, you must be almost single-mindedly devoted to your research. Being intelligent isn't enough—you must put in eighty-hour weeks poring over data and writing research proposals and grant applications.

I remember my father telling me about disputes he'd had with other scientists, explaining how many cling fast to their pet theories. While ideally a good scientist would remain open-minded, with no "preferred" result for an experiment, in reality, most don't. It's difficult for scientists to not be biased, hoping that their research will prove their theories. Challenging some "preferred" results, my father became well known within the field of birdsong science for disproving a commonly held theory about how a number of birds learned their songs.

As my father's challenge demonstrated, an established "dogma" can turn out to be wrong, and a good scientist is always skeptical. His success also demonstrates an incentive many scientists share: if they can derail a well-held theory, and develop one that better explains the evidence, they will become successful within their fields. But one thing is essential: scientists must always use data and facts to back up their arguments; theories alone are never sufficient.

This respect for the scientific method helped convince me that climate change is definitely happening. If a scientist could show that carbon dioxide doesn't warm the planet, he or she would become famous. Because of this, many have tried, but no plausible theory has been put forth to demonstrate that carbon dioxide won't increase global temperatures. Furthermore, the level of scientific consensus on climate change is almost unprecedented. Nearly every scientific organization, including the American Geophysical Union, the American Association for the Advancement of Science, the European Academy of Sciences, the U.S. National Academy of Sciences, and countless others (the list could fill a page) have all endorsed statements that humans are causing the Earth to warm.

Surrounded by five airport security guards in blue shirts and black pants, my father opened his suitcase, revealing a red bicycle folded over on itself. The bike was neatly arranged, with yellow protective sleeves over each tube. Dad carefully removed the parts and slowly assembled them. It was a special folding bike, with twenty-inch wheels, making it look like an oversized child's toy. Two even smaller wheels attached to the suitcase, which in turn attached to the rear of the bicycle, creating a trailer. Pops then unpacked his duffel bag and transferred the items to this trailer.

"I bought the most ridiculous looking bike I could, also just to embarrass you," he said, laughing.

"Thanks again, Dad," I said, secretly envious of how his bike packed.

The security guards just stood and watched as the gray-haired gringo assembled his bicycle in the airport lobby, and then biked out onto the street with his son.

We stayed with the friend of a friend in Managua, and then biked half a day to the colonial city of Granada, where, instead of trying to camp in someone's backyard or finding a fire station, my father opted for a hotel with air-conditioning. Pops, nearly sixty, could still bike all day long, but at the end of the day he needed a hotel room and comfortable bed. So, instead of a two-dollar meal of just beans and rice, deprived of vegetables on account of unclean tap water, we purchased an eight-dollar meal that included a salad washed by purified water. It was luxurious. On the other hand, I was now experiencing the country as a tourist would, and I no longer met firefighters in cities or farmers in the countryside.

My father had brought with him a number of important items to help me resupply: new high quality flat-resistant tires, a new pair of bike shorts, a bike shirt to replace my faded blue jersey, more business cards, silicone sealant to fix leaks in my tent, and a new LED headlamp to replace the one I gave Melvin in Honduras. The reasons I was glad to have my father join me were many.

NICARAGUA
April 13–20
Presentations: 0
Flat tires: 0
Miles: 299
Trip odometer: 5,361 miles

11 COSTA RICA 🚲
The Disappearing Mountaintops

Costa Rica population: 4.8 million
Annual per capita GDP: $13,000
Annual per capita CO_2 emissions from fossil fuels: 1.7 tons

DAY 168
April 21—Monteverde Cloud Forest

"I guess in Costa Rica people actually feed their dogs," panted my father as we biked up the hill.

Three small white-and-brown dogs yelped after Dad's trailer as we pedaled by an evenly cut lawn in front of a one-story house. I noticed the house was made of finished, painted wood instead of the adobe bricks or concrete of most Nicaraguan homes.

"Yeah," I replied, "the dogs in Nicaragua didn't look so well fed."

The dogs lost interest once we pedaled beyond the lawn. We then biked past lush rows of sugar cane, wet from a recent rain, as a modern green John Deere tractor sprayed pesticides or fertilizers onto the fields. When we passed a factory that processed the crop, the smell of cane filled the moist air. The rainy season, which comes earlier in southern Central America than farther north, was just beginning.

Unlike in Nicaragua, we saw no one-family plots of land, nor any men clearing the sides of the road with machetes. Instead, we passed a man using a gas-powered weed whacker. Costa Rica is the wealthiest Central American country, and while its average citizen makes less than one-third as much as the average American, the average Costa Rican is more than twice as wealthy as his or her counterpart in El Salvador, Guatemala, Nicaragua, or Honduras. The country's literacy rate is far higher than that of its neighbors as well. And while Costa Rica also had a civil war, it was more than half a century ago, and claimed few lives. Another advantage: the country's land has historically been more fairly distributed among its people than that of its neighbors, perhaps thus leading to less conflict between the lower and upper classes. Today, Costa Rica has the lowest rate of violent crime in the region.

———

We turned toward the mountains, passing signs reading VISIT US IN LA FORTUNA or TOURIST INFO CENTER. Every sign was written in English as well as Spanish—because Costa Rica is the safest and most developed Central American country, it also has the largest tourism industry.

We climbed, riding to the Monteverde Cloud Forest Reserve where my father had professional contacts from his research. The dirt road—a "short cut"—paid no heed to the terrain, climbing straight up the mountains at a 20 percent grade. Pops and I dismounted to push our bikes up the hill, and our feet slipped backward on the brown dirt.

Forty-five minutes before dark, we were still far short of Monteverde, with much climbing ahead. "I don't think I can make it," my father said.

"I'll wave you down a car."

"You don't have to do that." But as he said it, I hailed a pickup truck, which pulled over. The man in the truck agreed to give my father a ride up the hill.

As Pops and I lifted his bike into the back of the truck, he said, "Just don't tell anyone I got a ride."

"I won't, I promise," I replied, and biked up the hill on my own.

———

The following day my father and I followed our guide, Danilo, as he led a group of tourists through the Monteverde cloud forest. A Costa Rican standing six feet tall, Danilo walked backward as he addressed the group of tourists, mostly from the States.

"The diversity here is amazing," Danilo told us in near-perfect English. "Far greater than in northern latitudes. We have over fifty species of avocado trees here—they don't produce avocados you can eat; they're just in the same family. And that's just one family of trees!"

At 4,000 feet above sea level, the thick mist enveloped us—hence the name "cloud forest"—and at this elevation the air was a comfortable 65 degrees Fahrenheit. Because the forest is within the tropics, the temperature changes little over the course of the year, so it always feels like springtime in Monteverde. And the forest is aptly named: green covers most of the trees, as vines and moss grow on the bark, feeding on the mist.

"Here is a strangler fig. It's a tree that grows first as an epiphyte—a plant that grows on another plant. All the vines and mosses and orchids growing on these trees are epiphytes, and they survive because of the mist in the air, which carries both water and nutrients. The strangler fig starts growing as a vine attached to a tree branch. It then grows down to the forest floor where it extends roots into the ground. Eventually the vines encircle the trunk and become strong enough to support themselves. The fig's leaves spread above the original tree, shading and killing the host, and then the strangler stands on its own as the old tree rots away."

"Why don't strangler figs take over the forest?" asked a tourist.

"Well, not all succeed in growing large enough. Also, there's so much competition from other species, no single plant can dominate." I tried to take a picture of the tree, but mist covered my camera's lens.

"Look here under this step," Danilo continued, pointing beneath the boardwalk. "That's an orange knee tarantula." One by one we leaned over, looked under the wooden board, and with the aid of a small flashlight saw an oversized spider scared into the corner.

"Somewhere on this branch is an insect called a walking stick—can you find it?" We searched, twisting our necks to see the undersides of the leaves, but gave up. "Right here." He pointed to a long green insect blending in with the twigs right before our eyes.

We continued down the path. "Look—those are spider monkeys. They're much less common than the howler monkeys we heard earlier. But be nice to them! We had some tourists annoy them once and the monkeys started throwing their poop at us."

When a variety of birds leaped between branches my father and Danilo identified them. We saw a quetzal, a large green parrot-like bird, as well as various thrushes, wood-wrens, barbets, and a toucan. And my father's favorite: the three-wattled bellbird, a large white-and-chestnut-brown bird with strange "wattles" hanging off its beak. I was glad to hear its loud *bonk* that resonated throughout the forest, as Pops had recently helped prove that this bird learns rather than inherits its song.

Observing spider monkeys in the Monteverde Cloud Forest

Costa Rica has the largest network of biological reserves in Central America, and almost one-quarter of the country's land is protected like Monteverde. Just two of the many reasons for this: the successful tourist industry promotes conservation; and the country's low population density doesn't compete for additional land. But I found myself hearing again what Israel said at the National Institute of Ecology in Mexico City: wealthier countries deforest less than poorer countries.

This trend represents an idea within environmental economics known as the Environmental Kuznets Curve. The idea is that, as a poor country's economy grows, the nation initially degrades its environment; but, once the economy grows past a certain point, the country's citizens demand better environmental protection, and environmental degradation slows or even reverses. Thus, if one plots a graph with environmental degradation or pollution on the y-axis, and income per person on the x-axis, the Kuznets Curve looks like an upside-down U. For instance, while trash is prevalent along the roadsides of developing countries, it is not prevalent in richer nations. Rivers get polluted as a country industrializes, but then they are cleaned up as a country's economy grows. The Kuznets Curve can sometimes feel like a vast simplification, because other factors are also important: government corruption, distribution of wealth, the country's cultural values, even the fact that pollution can be exported. For instance, while the United States may protect its forests, it also imports trees from poorer nations. Nonetheless, the general trend—that pollution eventually decreases if a country's citizens become wealthy enough—appears to hold true for many types of pollution.

Some evidence suggests that Latin America is increasingly protecting its forests, as deforestation rates across the region are lower than they were a decade earlier. This trend is not a reason to celebrate, though, as deforestation rates are still problematic. This trend is also no reason for community complacency, since it's the very fact that citizens demand change that deforestation decreases. Nonetheless, the trend does suggest deforestation is a problem we might—with effort—solve.

The bad news, though, is that even if deforestation follows the Kuznets Curve—decreasing as incomes rise, and thus lowering a region's carbon dioxide pollution—fossil fuel use is expected to grow in step with those increased incomes, negating any possible benefit to the environment. And, while fossil fuel pollution may decrease once societies become rich enough to afford zero-emission technologies like solar panels, because carbon dioxide accumulates in the atmosphere, by the time we collectively reach that wealth it might be too late.

DAY 172
April 25—Monteverde's Mist

The night after our tour with Danilo, I attended a talk by Dr. Alan Pounds for a group of students with the Organization for Tropical Studies, at the Monteverde Biological Station. Dr. Pounds, a United States citizen, lived near the Monteverde reserve and studied the effects of climate change on amphibians. In 1988, the golden toad, a species found only in the Monteverde area, disappeared in what some say was the first documented case of extinction likely due to global warming. The toad relies on pools of water in the early rainy season to breed, and a series of dry years led to the toad's demise. Using weather data, Dr. Pounds showed that the region became drier because of warmer ocean temperatures; the warm oceans caused clouds to form at higher altitudes, reducing the mist and moisture in the cloud forest. While some more recent research disputes that climate change is responsible for the golden toad's demise, there is general agreement that a changing climate makes such extinctions more likely.

At the talk Dr. Pounds, a clean-shaven middle-aged man with glasses, outlined his recent research. "In one single genus of toads, *Atelopus,* so many species have gone extinct over the past twenty years. Of the original 110 species, all of which were found in Central and South America, only 36 remain." Dr. Pounds displayed a graph from his publication showing how almost all of the extinctions have followed unusually warm years.

Pounds explained that the climate is now more favorable to a fungus that preys on toads, and most of the extinctions appear to be driven by outbreaks of this fungus. "We can't account for this extinction just through habitat loss or invasive species," Dr. Pounds said. "This fungus is the best explanation." While I was somewhat skeptical of his fungus-climate explanation, the correlation between warm years and extinctions certainly was striking.

The following day I met with one of the many authors of that study, Dr. Karen Masters, at her ongoing experiment just behind the Monteverde Biological Station. Dr. Masters, who was probably in her late forties, stood in the mist in jeans, sneakers, and a light blue rain jacket, her hair pulled back into a ponytail. On my arrival she turned to face me, peering at me through her mist-covered glasses. She stood in front of a series of suspended horizontal logs covered with labeled tags and tiny plants, holding a spray bottle in her right hand.

After greeting me Dr. Masters pointed out the various logs with orchids growing on them. "We're testing to see how the change in mist affects the orchids that live here. As you know, since the 1970s the number of misty days in a year has decreased significantly. So, I have a few species here. Half of the plants we do nothing to, and the other half I spray with mist in just the exact amount to simulate what the conditions were like a few decades ago."

"So . . . you walk out here a few times a day and spray the plants with mist?"

"Well . . . you know that science isn't always that exciting! I have help. It's not just me."

"You must really like these plants."

She laughed. "I do, I really do. Anyway, after a few weeks of spraying, we measure various characteristics of the plants' size and health. On this row are plants that don't receive extra mist, and here are plants that do."

"These plants that you haven't sprayed . . . they don't look like . . . well . . . there doesn't look like there are any plants here."

"Yes. A number of them died. It is too bad—we lost them."

"I'd say that's pretty significant data! They didn't even make it. How do these species survive in the wild?"

"Farther up the mountain are places that receive more mist—but, of course, if the clouds keep rising, the species probably won't survive up there either."

"So, do these epiphytes have a future?"

"We don't want to jump to the conclusion that they'll all die. While mist is decreasing, cloudiness has actually *increased*—so there are more clouds, but they form higher up and don't deliver as much mist. More clouds means less direct sunlight and thus less evaporation, so maybe the plants can survive without as much mist."

"But . . . those plants on your log are dead."

"Yes, well, that's a good point," and we stared at the plants for a few moments. "The mist is what makes this place unique," she said, "and that could be at risk. The diversity here is amazing. We have about five hundred different species of orchids in this forest, many of which are found only in Monteverde. The ecosystem here is like an island at the top of the mountains. If the climate warms, these species won't climb higher because there's nowhere to go—they'd just go extinct."

"The climate has changed in the past, the Earth oscillating between ice ages and warm periods like the modern day. How have these species survived in the past?" I asked.

"Well, two things make this warming different. The first is that if you look into the past few hundred thousand years of Earth's history, it's never been warmer than we will likely make the planet. It's been colder, but never warmer. So, if you envision species moving up mountainsides as the Earth warms, and down mountainsides as the Earth cools, then we will move these species farther up the mountains than at any point in the past million years, and species at the mountaintops will disappear. The other reason is the rate of warming. A warming of 5 degrees Fahrenheit this century would be one hundred times faster than natural climate change."

We stood in silence for a few seconds. Karen looked at me and added, "What you're doing is so important—you need to go out there and tell as many people as possible. Losing this diversity would be a tragedy. It's like burning a library. Each species is a rich book, and we're destroying them. Once they're gone, they're gone."

DAY 173
April 26—San José

The following day, Dad and I bounced down another dirt road, heading toward the Pacific coast. At some point the sun emerged from behind the clouds, and the dry road dusted our bikes as we rode. I looked back toward the Monteverde cloud forest. A white cloud sat atop the green mountain like icing on a cake. Beneath the clouds, forest extended to lower elevations, but then the terrain gave way to land deforested for cattle grazing. The forest looked so fragile, trapped between the ranches below and the sky above. Even if Costa Rica entirely stops deforesting and fully protects its forests, as it may very well do, climate change could eventually rob the country of its diversity.

My father and I biked to the coast, where temperatures rose unbearably. Then, after a night in an air-conditioned hotel, we climbed back into the mountains to Costa Rica's capital, San José. There we stayed with the family of a friend, and I spoke at a local middle school and did an interview with Costa Rica's national television station.

Dad was soon heading back home, ending our two weeks of biking together. Before he left we discussed the road ahead, and I promised him I would avoid dangerous regions and choose my roads carefully. I gave him one of my four panniers, discarding items I wouldn't need in the warm tropics. I would later get a pannier mailed back to me when I reached colder stretches of South America—the high Andes and

Patagonia—but, until then, I would ride lighter and faster. Pops biked to the airport and returned to Massachusetts, leaving me again alone.

I couldn't help but appreciate my luck in being able to continue my journey. When my father was my age he already had a family, and worried about his career. By being born a generation later, I have had so many more opportunities for international travel than he ever had. Our society is wealthier now, and the reduced civil conflict in Latin America creates a far safer terrain for a globetrotting cyclist. When my dad discovered bike touring, his first instinct was to cross the United States; when I discovered it in turn, my urge was to ride to the far end of the world.

COSTA RICA
April 21–May 2
Presentations: 1
Flat tires: 1
Miles: 404
Trip odometer: 5,765 miles

12 PANAMA
Linked by a Canal

Panama population: 3.8 million
Annual per capita GDP: $16,000
Annual per capita CO_2 emissions from fossil fuels: 2.6 tons

DAY 180
May 3—Western Panama

I logged long hot days along Panama's only highway, a low-traffic two-lane road with a wide shoulder. Panama claims just over 3.5 million residents, a third of whom live near the capital city. While much land had been cleared for cattle grazing, the roadsides were as free of people as any land I'd seen since Belize. The air was thick and heavy, as if I were nearing the source of the tropics' humidity, and on every small hill the moist air weighed on me, slowing my progress. At night, to stay cool I slept on top of my sleeping bag and wore only my boxers. During the day, I sought out roadside restaurants with fans to cool me as I ate rice and beans.

Farther down the road, in Panama City, I would have a decision to make: how would I get to South America? No road connects Central and South America. I was shocked when I learned this; I had expected a heavily traveled road across the land bridge that links North and South America. But, alas, Panama's highway peters out after Panama City, giving way to the jungle and undeveloped wetlands of the fabled Darién Gap. I could potentially ride the few paths that do cross this border, but now that the rainy season had begun the trails were likely to be impassable mud.

A major reason no main road connects the two regions concerns Panama's desire to keep out the drugs and violence of Colombia. Guerilla fighters and paramilitary forces currently occupy the Colombian border with Panama, and these armed bands occasionally battle with one another, sometimes kidnapping tourists for ransom. I had recently heard of two British backpackers whose attempt to hike across the border landed them an unexpected nine-month stint as hostages in a remote jungle village.

Needless to say, I planned to skip the Darién Gap.

I decided to try to find a boat to the Colombian city of Cartagena. I wasn't yet sure if I would then bike across Colombia—a country with a forty-year ongoing armed conflict—or if I would catch a flight from Cartagena to Venezuela. I would make that decision once I made it to Colombia; first I had to cross Panama.

After a sweaty day of travel and a few occasional rain showers, I turned *del Fuego* off the highway toward the beach. Lining the road were a golf course and a set of apartments, with palm trees rising in neat rows as if transplanted from a retirement home in Florida. The apartments overlooked the Pacific on a bluff above a sandy beach; on the beach's edge stood wooden shacks with tin corrugated roofs.

Having ridden out onto the sand I lay *del Fuego* on the beach near the water, took off my shirt, and jumped into the waves, letting the saltwater wash off the day's sweat and grime. The Pacific was calm and warm. A bit later, resting on the beach, I chatted with a half-drunk civil engineer from Panama City as the sun set.

When it was almost dark I set out to find a safe place to camp—but I found my rear wheel was flat again, my third flat of the day. Flats always seemed to come in clusters. I'd had four flats in all of Mexico, yet in the two days in Panama I'd had five. Perhaps the new tires my father had brought were not as durable as I'd thought. I walked over to one of the tiny wooden shacks to ask if I could sit in a chair while I patched my tube.

"Yes, go ahead," said a man who was perhaps my age. He talked with me while I fixed my flat. As I pried the tire off the wheel, dirtying my hands, he told me that his family worked making and selling jewelry and souvenirs on the beach. They had three wood houses, each large enough for only two rooms. He suggested I set my tent between the structures—which sounded far safer than biking off into the dark road. But there was also a part of me that feared someone had purposefully punctured my tire while I was swimming, and planned to rob me later on.

Later, as I cooked my all-too-typical meal of pasta and cheese, family members filed out of the houses to watch me. So many emerged that it seemed like a magic trick. How did fifteen people fit in those three tiny buildings? The men stood with their shirts off in the night-time heat. I took a picture of us all together and wrote down most of their names: Edwin, Orbic, Amando, Abimelec, Tris, Alcide, Magheleen, Andy, Bélgica, Beatriz, Jaime, Siayeli. I noticed the family were of more African descent than Spanish or indigenous, embodying the ever-shifting ethnicities of Central America.

I let Andy and Alcide, who were maybe fourteen and ten years old, ride the repaired *del Fuego*. One rode on the rear rack as the other pedaled laps on a road that climbed the bluff and then descended to the water. "Ay yay yayaya!!" yelled Alcide as they raced down the hill, navigating by the light of infrequent streetlights.

Family I stayed with on Panama's coast

One of the men in his twenties pointed to Beatriz and said, "She is twenty-four and has no husband. You could marry her."

"I have to keep riding," I laughed, "and can't stay."

"Well, you could at least give her a child."

"But I couldn't help such a child. I need to keep riding."

"That's okay, we could email you pictures."

This exchange—and others—led me to wondering how different this journey would have been if I were a woman traveling by myself on a bicycle. While some say it would be impossible, I know of women who have done it alone. But such a journey would be far more dangerous and difficult, since women are treated far differently. For a woman traveler, offers to have a child would not be so innocuous.

DAY 184
May 7—Panama City

Another day of riding brought me to Panama City, whose cityscape immediately surprised me. Situated two miles from the canal, the capital city had the tallest, densest, and most modern buildings I had seen since Mexico City. Many were constructed on land created by filling in part of the bay, their bases only a few feet above sea level.

If sea levels rise as many project they will, in fifty or a hundred years this area will need seawalls for protection. This city *might* be able to afford such walls, as commerce and financial services have made Panama almost as wealthy as Costa Rica. But few countries can afford to build walls along entire coastlines, and families like the one I'd just stayed with would have to move. If the high-tide mark rises three feet, which it might do later this century, their land will likely be engulfed by the ocean.

Panama City lies on the Pacific coast. But I couldn't reach Columbia via the Pacific; few boats travel those waters, as they are supposedly filled with "pirates." So, since most boats travel to Colombia via the Caribbean, to continue by boat I needed to cross the isthmus. I could have easily done this via a forty-mile paved road, but I wanted to find a ride through the famous canal. Though I'd have to wait a few days or more until I found a sailboat willing to take me, I had a place to stay—with a friend of a friend who worked at the Smithsonian Institute in Panama. I also had a full agenda: talks at three schools, an interview with the national newspaper *La Prensa,* and updates to make on my website.

La Prensa put Ride for Climate on their front page, but I failed to get a copy of the article. To rectify my oversight, Susana, the woman who'd coordinated one of my recent talks, offered me a ride to her brother-in-law's office, as she knew he'd have one.

Susana's brother-in-law worked on the twentieth floor of one of the modern Panamanian buildings. When we entered his office we found him lunching in a small room with his three brothers and father, all wearing suits.

"Come, sit down, join us," said the father in perfect English, introducing himself as Jorge. Jorge, who was mostly bald, and had a thick mustache, was probably about sixty years old. He sat comfortably and confidently next to his sons by a window that offered a view across the city skyline to the serene Bay of Panama. As I sat and shared their meal, I vaguely recognized the runny beans and rice accompanied by thin tortillas. "Where is this food from?"

"Guatemala. We're all from Guatemala City," Jorge said, again in English. "It's the best city in Central America." His sons nodded in synchronous agreement.

"Unfortunately, Guatemala City is the only capital I didn't bike through. I was sad that I missed the city," I said, even though I had heard only negative stories about it: its crime, gangs, and run-down streets. Another cyclist had told me he'd tightly gripped a bottle of mace the entire time he was in the city's downtown.

When Jorge asked why I was in the paper, I handed him my business card and explained the purpose of my journey.

"Ah yes," Jorge replied, "that's a serious problem. You know, I gave a speech on global warming at the Rio Summit in 1992. I spoke just before President Bush." The Rio Earth Summit was the conference that resulted in the United Nations Framework Convention on Climate Change (UNFCCC), initiating the talks that created the Kyoto Protocol as well as the talks ongoing to this day. "Yes," he continued, "when I was president of Guatemala."

I wasn't sure how to respond to this information, and tried not to show my surprise. I wanted to say, *Wow—what's that like being president of Guatemala?* Or maybe, *I hear there's tons of corruption in Guatemala's*

*government; what was it like trying to run it? I mean, not to say you're cor-
rupt or anything . . .*

"What did you say in your speech?"

"I said that global warming was a serious problem, but that the poorer
nations should not be burdened with it. The wealthy countries created the
problem." Jorge then explained that the top priority in any poor nation
is to fight poverty, and such nations need to use the cheapest source of
energy they can. "The wealthier nations need to solve the problem."

"But don't you think we all have a role to play?"

"Yes, of course. But why shouldn't we be allowed to develop like
you have?"

He had a point: the average Guatemalan uses about one-twentieth
the amount of fossil fuel as the average American. But that doesn't
change the fact that the entire world is already polluting too much.
Again I felt a quandary.

After we left them, paper in hand, I asked Susana, "Why are Jorge
and his sons no longer in Guatemala?"

"He was expelled on charges of corruption."

"Were they true?"

"I don't know—we don't talk about that. But if they tried to return
to Guatemala, they'd likely be killed."

⸻

During my week off in Panama City, I thought again about what Jorge
had said about poorer countries deserving to develop as my country
had. I thought, too, about the family I'd stayed with on the beach. Current proposed international agreements on climate change would make
energy more expensive—whether through a direct tax or through a
"cap-and-trade" system. More expensive energy would create hardship
for the poorest of the poor, people who already struggled and suffered,
and who deserved the chance to seek a better life for themselves.

So, while updating my website, I created a page declaring what I
would do when I reached the tip of South America: fly home and bike
across the United States. This decision was driven by the success of
my outreach. I had visited a few dozen schools, and had placed countless stories in newspapers and television. My thinking was, if I could

actually influence people, I should continue the project in the United States, since my country—the second-largest polluter and the wealthiest nation in the world—has a vital role to play in solving climate change. As Jorge said, the poor countries will rightfully focus on developing, and they will use the cheapest energy they can.

After a week of posting notes on the Panama City marina bulletin board, I found a boat willing to take me through the canal. In exchange for working as one of four linesmen—who help with the ropes that stabilize the boats in the canal's locks—I would ride through the Panama Canal. Ray Durkee, an American sailing solo from California to Maine, offered me a spot on his thirty-seven-foot yacht, the *Velera*.

DAY 192
May 15—Panama Canal

The morning of departure, Ray listened to his radio and waited for the command to proceed. The *Velera*, a sleek white vessel with a white mast, prepared to make the passing through the canal. The water was calm, little wind blew, and the blue sky above had only a few wispy clouds. The three other linesmen joining us spoke no English, so I helped translate between them and Ray. Of course, they also knew how to tie the correct knots, whereas I was still practicing the basics on the deck of the boat. Though I had somehow convinced Ray I'd be able to help him through the canal's locks, I hadn't yet convinced myself.

Once Ray received clearance to proceed, he turned on the engine to power the craft through the canal. He then raised the anchor and secured it to the bow, and we motored toward what looked like a long skinny bay. On the right, a series of seven cranes rising over two hundred feet in the air lifted tractor-trailer-sized containers off a giant ship, then stacked them like Legos.

Ray, a man in his fifties, had a thin, fit frame and sported a gray beard and red baseball cap. "I've been dreaming of this day for a long time," he told me. "Passing through the canal is special."

As we motored toward the first lock, Ray shared the history of the canal with me. In the nineteenth century Panama was a province of Colombia; at the turn of the twentieth century, President Teddy

Roosevelt tried to negotiate with the Colombians to build a canal. After they failed to reach an agreement, the United States told Panamanian rebels that, if they revolted and succeeded, the States would support them. Soon after Panama declared independence, and the U.S. Navy prevented Colombia from reclaiming the territory. Immediately thereafter the U.S. began construction, finishing the canal in 1914. Rather than cutting a sea level waterway—which would have required digging a much more difficult and deeper canal—engineers instead built first a reservoir in the center of the isthmus, and then a series of locks on both sides, to raise and lower boats to the reservoir. "There's really nothing else like it in the world," Ray said. "And the engineering is amazing. They literally moved a mountain—you'll see the rock cuts once we rise through the locks."

"The canal used to belong to the U.S., but now it's Panamanian?" I asked.

"Yes, as of 1999, the Panamanians gained control of the canal. The U.S. used to control the land five miles to either side, which is why all the land remains undeveloped jungle."

We approached the first set of locks, where huge doors opened into what looked like a bathtub one thousand feet long. Each lock raises or lowers boats by twenty-five feet, such that the three sets of locks raise boats to the central reservoir, seventy-five feet above the Pacific. In front of us, a maybe seventy-foot-tall box-like ship slowly squeezed into the lock, clearing the sides by only a few feet. The back of the solid gray boat read HAUL AMERICA, and had what appeared to be a folded ramp. "Probably shipping cars," Ray remarked. The seven-hundred-foot boat gave us ample space behind it to maneuver in the lock. I felt like a toy duck in a bathtub.

"You have to be ready for the rope now," Ray said nervously.

Men walking on the edges of the walls— twenty-five feet above and fifty feet ahead—threw ropes down to us line handlers. The rope *thwapped* across the deck of the boat. I grabbed its end while the men above secured the rope to the top of the wall. The four of us on the boat tied clove hitches to four corners of Ray's boat. I tied the rope incorrectly and had to try again. "Hurry up," Ray stammered. As I finally secured the rope, water bubbled up from underneath, mixing the fresh water of the reservoir with the ocean's salt water, creating turbulent eddies beneath us. The boat wobbled and rocked, but the walls of the lock stayed far away. Slowly, the water level rose, lifting us to the next level.

After two more locks, we motored through the canal alongside the *Haul America,* traveling like a minnow next to a whale. Undeveloped rainforest lined the canal, which occasionally exposed rock where the mountains had been blasted away. Heading the other direction, a container ship passed us; loaded on its deck were stacked a few hundred tractor-trailer-sized boxes. I wondered how many goods I've owned—how many T-shirts, toys, electrical devices—traveled this waterway. Nearly fourteen thousand ships pass through the canal every year, accounting for about 5 percent of all world trade.

During my time in Panama City I'd researched the environmental impact of these ships, and I was impressed by their efficiency: per mile it's fifteen times more efficient to ship by boat than by truck. So, though these ships represent a huge portion of world trade, they produce only a small fraction of carbon dioxide pollution from transportation, and are responsible for only a little more than 1 percent of all greenhouse gas pollution.

In fact, shipping statistics reveal an interesting truth: given that the shipping of goods is actually extremely efficient, buying local has only a small effect on one's carbon footprint. A study by researchers at Carnegie Mellon University showed that only about 10 percent of the carbon emissions from food derives from its transportation. The majority comes from growing and processing the food; a farm's efficiency is actually more important than its location. Another study showed that, because sheep are raised much more efficiently in New Zealand than in England, British citizens' carbon footprint would be smaller if they ate mutton imported from the far side of the Earth rather than from the local farm.

One reason the transit of goods is efficient is that ships and trucks are packed full of items—a large truck can carry forty thousand pounds of goods. By comparison the family car, which weighs a few tons, usually carries only one or two people, and burns ample gas in order to carry just a few hundred pounds. It turns out that driving less or in a more-efficient vehicle almost always has a much larger effect on reducing one's carbon footprint than does buying local goods. That doesn't mean don't buy local; it just means it's less important carbon-wise.

Nonetheless, even the more efficient ocean-traveling transportation creates pollution. Many such boats have inexcusably bad sulfur and particulate pollution, which creates horrible air quality in port cities. In the long run, we need to approach ships and trucks in the same way we approach the rest of the economy—by improving efficiency and searching for cleaner fuels.

But when I saw these floating giants, I didn't think of their pollution; I thought of how the world is truly connected. These ships symbolize that we are all in this together, buying and selling goods around the world. World trade has exploded in recent years, doubling in the decade before my journey. I hope that the world's increasing interdependence will encourage us to solve problems like climate change, problems that have no borders.

Container ship behind us in a Panama Canal lock

The canal opened up into the reservoir in the middle of the isthmus, Gatun Lake. Tropical rainforest lined the shoreline, with not a building in sight. We passed perhaps a dozen ships cruising through the lake, some a thousand feet long. A yellow-hulled rusted oil tanker rode low in the water, with another container ship, this one with a red hull, cruising behind it. This ship was in turn followed by a boat nearly identical to the *Haul America*, but turquoise instead of gray. I felt as if we had stumbled across the Lost World, though one of diesel giants rather than of dinosaurs.

After descending three locks on the far side of the reservoir, we arrived in a bay that opened to the Caribbean Sea. In the day's fading light, we saw a fleet of large ships on the horizon, all anchored and waiting to pass westward through the canal. Ray opened a cooler and handed each of us an Atlas beer—the Panamanian beer—and toasted our success. We dropped anchor near the local marina, and a small motor boat took me and the other linesmen to shore, where I would look for a boat headed to Colombia.

DAY 198
May 21—Colón to Cartagena

After giving a presentation at a local school and spending a few days at the marina in Colón, Panama, I met Peter, a skinny Austrian with skin tanned and dreadlocks bleached by the sun. He owned a sixty-five-foot luxury yacht, pretentiously named *The Golden Eagle*. He would soon make the voyage to Cartagena, and was looking to fill the boat with twelve paying customers. As I was at the marina a few days before departure, I helped Peter sand his boat's wooden floors. In exchange, Peter decided I could be his "first mate" for the voyage, a status that appeared to give me the right to go on *The Golden Eagle* a day early and sand its floors.

The next day we were joined by eleven other passengers, all foreigners in their twenties or early thirties, traveling around Latin America carrying nothing more than a backpack: two friends from England, a woman from the Netherlands traveling with them, a twenty-five-year-old man from South Africa, a Canadian woman just out of college, two Australian friends traveling with their surfboards, and an attractive woman from California who, oddly enough, once lived in the same dorm as I did in college but who was for some reason entirely uninterested in me and far more interested in the Australians.

While I have enjoyed traveling with others in the past, I noted that this time I didn't feel I belonged with non-cycling travelers. Our experiences were so different. When traveling without a bike, one regularly climbs onto buses and piles into hostels with other travelers. On *del Fuego*, I rarely saw other foreign tourists. After so many months traveling mostly on my own, communicating almost entirely with people in new lands, I felt like a social misfit, unable to communicate with my "own people." I almost wanted to be back at a fire station. At least *del Fuego* passed the journey with new friends; the bike rode hidden in a large plastic bag, tied to the deck alongside the two Australians' surfboards.

San Blas Islands, off the coast of Panama

After anchoring for a day by the San Blas Archipelago, a series of palm tree–covered islands that looked like they belonged in a *Far Side* cartoon, we sailed east and left Panama for good. I was excited to move on. With its hot temperatures and frequent borders, Central America had exhausted me.

From the deck of *The Golden Eagle,* watching the Panamanian shoreline disappear behind the curve of the Earth, I thought back on my time crossing Central America. I had witnessed so much that was at risk to climate change, including reefs in Belize, farms in Honduras, cloud forests in Costa Rica, and coastlines in Panama. It made me discouraged to contemplate how, if we are to save such incredible, beautiful, and diverse environments, we must act now, and on a large scale.

Yet, despite such thoughts, my most visceral impressions from the region were not of climate change, but of poverty, conflict, and corruption—Melvin living off the land, struggling to make ends meet with no alternative opportunities; the stories of Delmy in El Salvador and war's capacity for destruction and long-term suffering; the former Guatemalan president in exile, likely guilty of corruption charges. Being able to worry about climate change almost feels like a luxury compared to worrying about war and starvation. I only hope that the solutions to these problems aren't mutually exclusive. Why can't we have a peaceful, prosperous civilization that doesn't increase the Earth's temperature? That must be possible.

After two more days sailing and battling seasickness in the open ocean, we watched the buildings of Cartagena, Colombia, rise up from the ocean. South America lay ahead, a continent I planned to cross from north to south, traversing 65 degrees of latitude. I was eager for my next adventure, and each country on my route fascinated me. Colombia struggled with armed conflict and was ruled by a conservative hardline president. Venezuela managed (or mismanaged) huge oil riches and was led by a freewheeling left-wing president. Brazil boasted the Amazon jungle and river; and to the west, in Peru and Bolivia, the Andes mountains and Incan ruins awaited. In northern Chile and southern Bolivia, the Atacama Desert had produced the world's largest salt flats, many of which can be biked across. I planned to visit all of these places before riding south through Chile and Argentina, following the Andes like a line on a treasure map to the end of the road in Tierra del Fuego. But, first, I had to cross Colombia.

PANAMA
May 3–24
Presentations: 5
Flat tires: 10
Miles: 344
Trip odometer: 6,109 miles

PART III
COLOMBIA AND VENEZUELA 🚲
Bikes & Oil

When I crossed into Colombia, my home finally felt like a distant memory. Much as the sail from Baja California to mainland Mexico had made me feel my adventure in Mexico had begun, this longer sail from Central to South America made me feel I had completely lost myself in the journey. Even with California many months behind me, the end of my route was still at the other end of the continent; I had plenty of adventure still ahead of me. I'd also be traveling "off route"—going south, north, and then east, wandering around the northern end of the continent—before biking south into the Amazon, as if to emphasize the fact that I was in no hurry to reach my goal.

My explorations through northern South America taught me some of the most fascinating lessons of my journey. For one, the political situations of Colombia and Venezuela surprised me; though the two countries butt against each other, they seemed to be traveling in opposite directions. And while I had expected to enjoy Colombia's biking culture and dislike Venezuela's oil economy, I had no idea how strongly I'd feel these emotions, or how illuminating it would be to hear Colombians and Venezuelans discuss their countries' circumstances. I continued to feel amazed that my bicycle journey gave me such an intimate window into such diverse situations, and into so many different people's lives.

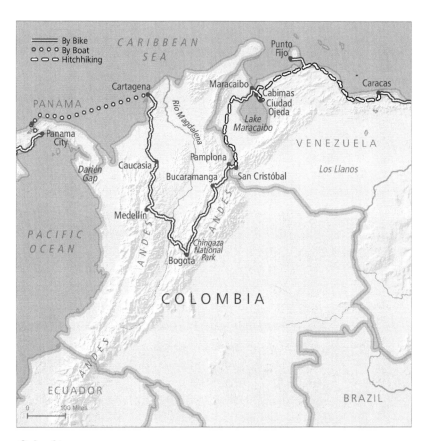

Colombia route

13 COLOMBIA
Cities Reborn

Colombia population: 48 million
Annual per capita GDP: $10,600
Annual per capita CO_2 emissions from fossil fuels: 1.6 tons

DAY 205
May 28—Arjona

"You were kidnapped by the FARC?!" I asked in shock of my host, Manuel. We sat in the single-story home of his parents on a hog farm in Arjona, a town twenty miles inland from the Caribbean. The house, though smaller than most U.S. homes, was comfortable and included a somewhat modern kitchen with a gas stove and refrigerator. But the screen windows failed to let through any breeze, so I sweated in the heat and humidity.

"Yes," he replied, unfazed by my surprise. Manuel, in his late twenties, had smooth skin and dark curly hair. "Three years ago, while taking a cheese shipment to Medellín, I was pulled over by the FARC." He explained that they held him for three days until his parents paid five thousand dollars. "It was terrifying—I was blindfolded the entire time. I thought I was going to die."

I almost didn't visit Colombia. Like many Latin American countries, Colombia has endured a violent struggle between Marxist guerilla fighters and the government. But in Colombia the struggle continues, and large swathes of the country are not safe for tourists, largely because the guerilla fighters kidnap foreigners for ransom. The 2000 film *Proof of Life* is more or less based on the situation in Colombia, although the story takes place in the fictional South American country of "Tecala." In the movie, actor David Morse is abducted by Marxist guerillas and held hostage in the jungle for more than half a million dollars. He tries to escape, but fails and is tortured further by the rebels. Eventually Russell Crowe lands with a helicopter and rescues Morse, with the assistance of explosions and a deluge of bullets. I did not want to risk a similar fate. I didn't want to bankrupt my family, and I doubted Russell Crowe would rescue me.

In trying to decide whether to cross the non-fictitious country of Colombia—and lured by stories of warm people and a bicycling culture—I had researched the country's present-day reality for several months. The Colombian conflict has never been intense enough to be called a civil war, and is instead referred to as an armed conflict. Nonetheless, this quasi-war has continued for forty years, killing somewhere between fifty thousand and two hundred thousand people. Two separate organizations, both of whom claim socialist aims, fight the government: the FARC (Fuerzas Armadas Revolucionarias de Colombia) and the ELN (Ejército de Liberación Nacional), the FARC being the larger and better-known organization. The government has historically failed to quell these armed conflicts, and in response private right-wing paramilitaries have sprung up throughout the countryside in armed response against the FARC and ELN. The paramilitaries sometimes receive direct support from the government, but, because they are not an official part *of* the government, they can respond with greater violence and greater disregard for human rights. Fortunately, within the previous several years many paramilitary groups have disbanded under a government "peace and justice" program, and both the FARC and ELN are decreasing in strength. But remnant paramilitaries, the FARC, and ELN still run de facto local governments, fight for territory, and fund themselves through cocaine trafficking. And, as my host could attest, the FARC and ELN still kidnap Colombians and the occasional foreign tourist for ransom. From what I've heard, these groups mostly seek money, not political gain; many Colombians told me the conflict "used to be about ideology. Now it's about money."

In the years before my visit safety had dramatically improved, partly due to aggressive offensives by the Colombian army. The FARC and ELN have shrunk in influence, and numerous paramilitary armies have disbanded. Nonetheless, the country still had dangerous regions controlled by guerillas, and perhaps 20 percent of Colombia—a nation the size of Texas and California combined—is still in the hands of the ELN, FARC, or paramilitaries.

So, in planning my journey through this terrain, I emailed numerous travelers to determine a safe route, finally connecting with a U.S. official who studied the conflict and frequently traveled to Colombia. After looking over my itinerary, he wrote and agreed my route would avoid guerilla strongholds. "I say go for it. Just stay in towns at night, stick to the main road, and you should be fine." Fine? In my first week in Colombia, I sat in the room of someone who'd been kidnapped by the FARC.

I had met my host by chance, when Manuel pulled up alongside me in a pickup truck as I biked out of the coastal city of Cartagena. He asked if I needed a place to stay that night. Wary of accepting help from strangers—and fearful of, say, being kidnapped by the FARC—I hesitated, then asked him a series of detailed questions. Where did he want to take me? How far away was it? What did his family do? Why did he want to help me? His calm answers, and my gut, which was the only guide I really had, told me I could trust him, that he was just excited to help a foreign traveler. So I wrote down directions to his home, and followed him there; and there was where I now sat, learning about the FARC.

"So," I said to Manuel, "your parents paid five thousand dollars and the FARC let you go?"

"Yes, that's how the FARC makes their money, along with selling cocaine," he replied.

"But, *you think it will be okay for me to bike across the country?* As an American? Why is it so much safer now?"

"Only Uribe has the guts to stand up to the FARC," he said. "The military is 50 percent larger than it was five years ago, and it's now safe to drive to Bogotá."

"Is it safe to leave the roads?"

"Often not, but the roads are now safe. Thank God for President Uribe."

In almost the next sentence, though, Manuel advised me about dangerous roads ahead. "When you bike to Sincelejo, I recommend taking this road and not this one," he said, pointing to the map I'd laid on the table. "This here is paramilitary territory. Actually, Sincelejo, where you'll be tomorrow, is more or less run by the paramilitaries."

"I thought you said it was safe!"

"Yes; the paramilitaries usually don't bother tourists. You'll be fine, but this route is safer." He then described how government officials had recently uncovered a mass grave outside of Sincelejo.

DAY 208
May 31—Caucasia

As I traveled south, the landscape was an almost unnaturally lush green beneath the gray cloud-covered sky, as if I had turned up the "vivid" setting on my camera. The air held the smell of moist earth. Much of the rolling countryside was cleared for cattle or pigs, with the occasional lone, tall tree arching over the road. Traffic was light and consisted mostly of trucks transporting goods—only a few private cars passed me. At one point an open bed truck lumbered by full of men in camouflage uniforms. Each man held an M16; a few of them smiled and waved.

Farther along, a shirtless man on the roadside stood by a few cartons of mangos. He wore a faded New York Yankees hat and his skin clung tightly to his skinny frame. He smiled at me and said, "*¿De donde viene?* (Where do you come from?)"

"California," I replied.

"For you," he said, and passed me a mango.

As I rode, nearly every person I encountered looked at me and asked, "*¿De donde viene?*" It was always asked with a surprised expression and a smile of curiosity. The reaction was both welcoming and unsettling, as it suggested I was a rare sight—and that I was far from where foreigners travel.

———

In Caucasia, a medium-sized town along the banks of the wide Cauca River, I biked to a middle school and told the secretary about my project. It was a public school, built of concrete with white walls. I was soon directed to a class of seventh graders studying English.

Here I confronted the same challenge I'd previously felt visiting public schools. While I believed that such students should know more about global warming, I wasn't sure what they could do with the information, other than become angry with people like myself who were far more responsible for it than they were. To be honest, part of my motivation to visit the schools wasn't just to educate the students—it was because I was curious. I wanted to see how students learned, and what school was like in Colombia.

The English teacher was happy to take a break when I showed up. "Yes, you can talk to them," she said, and departed almost immediately, leaving me, without introduction, with a room of thirty seventh graders. The boys wore blue pants and the girls blue skirts; all wore white-collared tops. But the uniforms were the only orderly part of the class. The chairs and desks were randomly angled. The concrete floors loudly echoed the students' chatter, and even more noise poured in from the concrete courtyard outside, where other classes were at recess. A number of younger students pressed against the window to look at me and *del Fuego*—they were still young enough to not be embarrassed by curiosity. Some of the seventh graders in the class stared at me, trying to figure out how much interest to convey. A few yelled, in English, "Good morning teacher," as well as, "What is your name?" These statements seemed to be the limit of English education across much of Latin America. I met countless children who repeated these exact phrases to me, often struggling to pronounce the long "ay" sound of *name*.

To my question of if the class could count to ten, I got blank stares. So I asked in English, "What else do you know how to say in English?" A number of the girls laughed. I repeated the question; the room remained silent.

Reverting to Spanish, I explained how far I'd traveled, and momentarily got their attention. I asked who had a bicycle (about a quarter did), and told them they too could bike across South America. (I sometimes told the story of the Argentine cyclist who crossed Latin America with no money, relying entirely on the generosity of strangers.) I then asked how many students attend their school, and wrote down the number on the board. I then asked how many people live in Caucasia, in Colombia, and then the entire world, writing their replies in white chalk. I talked about what they share with these people. "You share this courtyard with your classmates, and if you throw trash in the courtyard, it's disrespectful to the other four hundred students at the school. You share the town's main plaza with the other people in Caucasia, and if you throw trash there, you're disrespecting them. We share the air with 6.5 billion other people around the world, and we need to be careful with it—if we contaminate it, we're disrespecting the people around the world."

I took out a large, partially ripped map of South America and Central America, and had the students gather round so I could point out my route.

"Where are we on this map?" I asked.

The students puzzled over the map. None pointed out Caucasia. I was surprised, but only somewhat. So often I'd show people along the roadside maps of the town or region where we were, just to find they didn't know what to make of the map, because they don't use them.

"We are right here, on the Cauca River. Does this river ever flood?" I asked, curious.

"Yes," said one of the boys. "About once a year. It's great—school is canceled."

"Why?"

"Because the school is flooded." A number of the girls laughed. The river ran right behind the building.

I again felt confused. Flooding no doubt takes a serious toll on their livelihoods and education, and most likely such floods will worsen with climate change. Yet the much more immediate challenge is the fact that the students are likely getting a sub-par education, in a town where there will be few opportunities when they graduate. And, even worse, there is an armed conflict in the hills surrounding their home. Why should they care about sharing the planet with the other 6.5— now 7—billion people on the planet, when there are so many more immediate threats to their livelihoods?

———

That night I found a hotel next to a gas station along the main road. Cooking a meal of pasta and canned tuna on the steps of the hotel, I noticed four men from the military standing in the parking lot. They wore full-body camouflaged gear, and each carried an M16 over his shoulder. One of the men, who was maybe twenty years old, approached me and said in English, "Hello Meester!"

"Hello." I considered correcting his pronunciation, or pointing out that we don't say "Hello, Mister" the same way they say "*Hola, Señor,*" but, as he held a very large gun, I decided his pronunciation was just fine.

"Where do you come from?" he asked in Spanish.

"I'm from California."

"Welcome to Colombia," he replied, and gave me a smile that appeared far too friendly and relaxed for a man with two grenades clipped to his belt. He introduced himself as Juan. After I explained my bike trip, Juan said, "Bush is a great president."

"Why do you say that?"

"Because he goes after the terrorists," Juan said. Juan then explained that the Bush administration had provided financial support to Colombia, which in turn allowed Colombia to expand its army.

"Have you ever fought the FARC?" I asked.

"Yes," Juan replied.

"Ever killed any of them?"

"Yes, I've shot four."

"Have you lost anyone in your platoon?"

"Yes."

"Wow. Well, thank you for securing the roads."

"No problem."

Juan then posed for a picture with my bike, and I asked for another picture with him pointing his gun at my pot of pasta on the ground. The other soldiers saw this, walked over, and asked for me to take their picture as well.

After eating, I carried *del Fuego* into a small eight-dollar hotel room with yellow painted walls and a ceiling fan. I was told not to camp in the countryside through this stretch of Colombia, so I had to find hotels when I was in towns too small to have fire stations. Lying on my back on the bed, I attempted to read the Colombian novel *One Hundred Years of Solitude*, by Gabriel García Márquez, in the native Spanish. While the book's language was beautiful, I only understood every other sentence, as my Spanish was only so advanced. I put the book down and stared at the rotating ceiling fan, and then looked around the room. To my right, the contents of my panniers had been strewn about. My pot lay on its side against the wall, drying after being washed in the sink; my biking shirt was draped on *del Fuego* after being washed in the shower. My map was half open on the floor and my sleeping pad was airing out. When I stayed at people's houses or at fire stations, I was always careful to not make a mess, always conscious of being in someone else's space. In a hotel, however, I could expand my items to occupy the entire room; clearly I'd taken advantage of that opportunity.

Yet I was filled with loneliness, aware that the hotel room intentionally bore no memory of past guests. In paying for privacy I'd also paid to have no one to talk to. When I camped behind a family's house

I often befriended them, and at fire stations the *bomberos* were always eager to tell me stories. Even when I set up my tent in the remote countryside I was rarely lonely—maybe because I felt somehow connected to the rest of the world while sleeping out on its landscape. But in a hotel I was alone, with walls caging me in.

Juan (at right) and other soldiers in the Colombian army aim at my dinner

In that Colombian hotel loneliness mixed with frustration. I was irritated by my route, even though it was of my own choosing. I had decided to bike to Bogotá, Colombia's capital, having been invited there by Ricardo, the bicycle advocate I'd briefly met in Mexico City. I'd also made a point of visiting every capital city I could along my journey, as such places are centers of culture and energy, not to mention the best places to get national news attention. But to get to Bogotá involved a long detour to the south, and then a long return north, making a giant V across the country. Bogotá was "off route," and there was something deeply unsatisfying about riding off route, as each mile I pedaled didn't get me closer to my final destination. Admittedly, my frustration probably also resulted from the heat, which exhausted me every afternoon since I'd reached the tropics.

I think I also felt unsettled, questioning the value of drawing attention to climate change. Or, rather, I was no longer sure how to talk about it. I knew that Columbia will probably regret a warming climate, even though the country has so many more pressing, immediate challenges. In the long run, rising oceans could eventually engulf the coastal city of Cartagena, which the Spanish founded in 1533, almost five hundred years ago. And if powerful storms displace communities—much as Hurricane Mitch did in Central America—the resulting chaos may feed into the hands of the FARC, ELN, and paramilitaries. Countries with civil strife always struggle after natural disasters, as humanitarian aid is difficult to deliver to areas controlled by guerilla armies.

But how should we compare these long-term issues with the need to improve livelihoods today? As I biked across Colombia, I was far more afraid of the FARC than I was of climate change. The FARC was a short-term threat—armed bandits could have done me in any day—while the threat of global warming at its worst will hit decades from now. But, then again, to avoid the worst of climate change we must act now, as it takes decades to replace our energy system. If we wait, the climate will get even worse—from which it will take longer to recover, with even more lives adversely affected in the meantime.

DAY 210
June 2—The Andes

After crossing the wide brown Cauca River, its sediment-laden currents draining recent heavy rains from the mountains, I began to climb. I had finally reached the Andes, and was at the northern extent of the world's longest and second-tallest mountain range. As I ascended their northern edge I felt the great ambition of this journey; within a year I hoped to roll down the southernmost end, and I envisioned the peaks guiding me southward. I dreamed of what lay at the far end, on the island of Tierra del Fuego—imagining glaciers, cold winds, penguins. But right then, in the Colombian Andes, I encountered jungle alternating with deep green fields cleared for agriculture. I sweated as I climbed, my biking jersey clinging to me. Nearly every mile I stopped and rested, waiting for my body to cool before continuing.

As the temperature dropped as I gained altitude, my legs found strength, and my general frustration subsided. The road cut into the wet hillside, and in places landslides had damaged the highway, reducing it to one lane. Construction workers in yellow hard hats with hand-powered cement mixers worked to fix the road. Above, wisps of fog extended from a cloudy ceiling. When a slow-moving truck passed I sprinted to catch it, hoping to hang on its rear and be pulled up the hill. Two boys, maybe in their late teens, clung to the rear of the truck, sitting on its back bumper. I tried to grab onto the vehicle, but it had no easy points to hold onto—no large latch or rope. When the taller of the two boys reached out to offer a hand, I extended my arm; he grabbed my wrist with his left hand, while his other arm securely gripped the truck.

"Are you sure you can hold on?" I asked him.

"Sure," he said. "Where do you come from?"

"California. And you?"

"We went to the coast for a month, looking for work. We're headed home to Medellín."

"Did you find work on the coast?"

"If we did, we would've paid for a bus instead of hitchhiking." He said this matter-of-factly as the truck groaned and rounded a turn. I struggled to hold my handlebars steady with my right hand while holding onto the boy's wrist with my left. After about a minute my arm tired. I let go, yelled thank you, and watched the two boys on the back of the truck disappear into the cool mist of the Andes' clouds.

———

After two days of cycling through the mountains, including ascending a 9,000-foot pass where a frigid, heavy rain pelted me and rainbows arched across the sky, I descended into the valley of Medellín. Colombia's second-largest city, with over two million residents, occupied a long green valley, with buildings creeping up its sides until the neighborhoods gave way to lush forest. I had heard that Medellín used to be the center of Colombia's cocaine trade, and had been known as "the murder capital of the world." It was also the home of Pablo Escobar, perhaps the most wealthy and infamous drug lord in history—until he was gunned down in the early 1990s.

As I rode into town on a Sunday morning, I found a road blocked off and filled with cyclists—the *ciclovía*. In major Colombian cities every Sunday, the principal streets are closed to cars, open only to bicycles and pedestrians.

As I cycled into the multitude a man maybe fifty years old wearing sweatpants and a calm smile rode alongside me on a simple mountain bike. Graying black hair poked out from under his white baseball cap. "Where do you come from?" he asked; I explained how I'd biked from California. Introducing himself as Lucho, he offered to give me a tour of the city.

Though I was skeptical, Lucho's smile made him appear too content to have anything up his sleeve. So he led me along the street, pointing out major landmarks. Most of the city's structures were made of brick, the poorer homes clustered together like stacked boxes lining the long valley. Although in the tropics, the city was a mile above sea level, and the comfortable temperature relaxed me. "It doesn't get much warmer or colder than this," Lucho said. "They say this is the city of the eternal spring."

We passed children riding small mountain bikes, adults on rusty one-speed cruisers, women in tight-fitting sweatpants, and men in Spandex riding new road bikes. We paused at a stand selling fruit next to where a man led a group in aerobics accompanied by techno music. Lucho bought me a cup of *salpicón*—mangos, pineapple, and watermelon with whipped cream. "It's on me," he said, though he couldn't have had too much money to spend; he told me he worked as a plumber, and his plain clothes and bicycle suggested he spoke truthfully.

We rode to a park surrounding the city's soccer stadium, where street performers played accordions and guitars. In another park, in front of a giant brick cathedral, an orchestra provided a live concert. Nearby, a garden was filled with rounded corpulent sculptures made by the Medellín artist Fernando Botero. We rode to another plaza where a priest gave an outdoor mass, alternating his sermon with lively songs, to which the standing congregation swayed together as they sang.

Then a heavy rainstorm suddenly passed through town, obscuring the sun and briefly flooding the streets with water. Retreating beneath a pedestrian bridge, Lucho and I watched children jump in an outdoor fountain and laugh in the rain.

"How does Medellín seem to you in comparison to how it's painted on the news?" Lucho asked me. "Isn't it beautiful?"

Bomberos de Medellín

"Yes."

"And aren't our women beautiful?"

"Ah, yes."

While it's true there were beautiful women in the city, it was the quality of the cycling that made the stronger impression. Having opened these roads to bicycles and not cars, the city felt more open, more welcoming. In cars we are trapped in boxes that separate us from the environment. Whereas on a bike, a man like Lucho was easily able to approach and befriend me.

At the end of the day I rode to the fire station, where the city's *bomberos* welcomed me and gave me a BOMBEROS DE MEDELLÍN jacket, uniform, and T-shirt. They also bragged to me about the beauty of the city and its women. I could imagine most Medellín citizens hoped I would return to the U.S. and belie the flawed news reports, telling everyone what a great city theirs is.

DAY 215
June 7—Magdalena River Valley

With my panniers bulging with Bomberos de Medellín paraphernalia, I climbed 2,000 feet out of the valley, riding east toward the capital of Bogotá. At a 7,000-foot pass, a cold, heavy rain forced me into the town of Santuario. When I asked directions for the road ahead, a woman at a small convenience store warned me, "Don't take that road at this hour. They close it at night; it's safe only during the day, when the army sets up patrols. You should stay in Santuario and then ride tomorrow." My plans thus determined for me, I thanked the woman and found a hotel room. In the morning and the safety of day, I followed the highway out of town.

The Andes in Colombia are not one uniform range, but rather three parallel mountain ranges. The cities of Medellín and Bogotá, both perched high in the mountains, are separated by the Magdalena River and its deep valley. To reach Bogotá I had to descend more than 7,000 feet, nearly to sea level, and then climb back up another 9,000 feet.

Riding toward the Magdalena River Valley, I found the roadside almost entirely devoid of people and settlements. Dense forest covered steep slopes above and below the road, and intermittent rainstorms drenched me as I descended and climbed a series of ridges. A sign read in Spanish, TRAVEL PEACEFULLY. YOUR ARMY IS ON THE ROAD. This claim was backed up by an occasional man in camouflage standing on the roadside with a machine gun. Though most of these soldiers waved and smiled, one uniformed man stopped me.

"Show me your papers," he said, in an accusing tone.

As I dug in my pannier for my passport, trying to stay calm, a second man in uniform approached the first. "Let him go," he said, "don't bother him," and waved me on. I continued, breathing a sigh of relief, and wondered if I'd just narrowly escaped disaster. Another cyclist had told me how, while traveling in southern Colombia, he'd been held at gunpoint and robbed of all his gear.

I descended into a steep valley whose walls were covered in deep green forest, the hillsides only occasionally cleared for agriculture. I then crossed a bridge over a mountain stream with bubbling rapids and climbed up the far side. This route was not one long descent to sea level, but rather a tortuous roller coaster of climbs and descents across steep valleys blanketed with rainforest. And even though my route connected the two largest cities in Colombia, little traffic passed me. I found myself not surprised that the central government didn't have control over the country; the mountains and jungles dividing Colombia made me wonder if it was possible to govern it at all.

Magdalena River Valley

At one point in my climb my pedals swung suddenly, giving no resistance, and my chain fell to the ground, snapped in half. After momentarily losing control of the bike, I walked *del Fuego* off the road, propped it on its kickstand, extracted my chain tool from my pannier, and began the tedious process of fixing the broken links.

I stood in front of half a dozen tiny houses—some of the scarce structures along this road. A woman in one of those houses walked to her doorway and stared at me. Her gaze was fearful, distrustful, unwelcoming, and she stood just far enough away so as to not initiate conversation. She also looked weathered by something other than time. I gave her a friendly nod; she didn't respond. I thought to myself, *On these roads, when the army disappears during the night, what happens? Has she lost family members in this conflict? Would I be kidnapped if I spent the night here?*

I struggled to reconnect the chain. What I'd thought would be an easy fix was not. I had to remove a broken link, slightly shortening the chain, and then force a pin back through to connect the links. My hands became completely covered with grease, and I also got marks on my arms and legs, sweat mixing with grime and oil. I wondered if the woman thought I was a wealthy foreigner. Nervousness crept through my body. I'd be stuck without a functioning bike. I'd have to ask for help, maybe even spend the night on this forbidden track of highway. Sweat dripped down my arms and over my greasy hands. After about twenty tiring minutes, the pin finally set, and the chain was whole. Relieved, I remounted *del Fuego* and continued up the hill, the woman still watching me.

After a seemingly endless series of climbs and descents, I reached the Magdalena River, a giant swath of brown water between the rounded verdant banks of a wide valley. Few sizeable cities lined the river in this stretch, the populations instead perched high in the mountains on either side. Civilization felt far away. Spent rainclouds crossed the sky, occasionally clearing to let sunlight fall on the green valley. I half expected to encounter Macondo, the mythical small town in *One Hundred Years of Solitude*, a town that had almost no contact with the outside world except for a band of gypsies that passed through once a year.

DAY 220
June 12—Bogotá

After three days of arduous climbing and staying at small hotels on steep mountainsides, I reached the plateau. It surprised me. At more than 9,000 feet, a flat twenty-mile-wide green landscape spread to the horizon. On the far end of the plateau, buttressed against taller mountains, was the city of Bogotá, Colombia's capital and home to seven million people. A lake had once filled the valley; the sediments it left behind created this flat terrain between mountain peaks.

I was joined at the valley's edge by Bruce, a U.S. citizen who worked for the U.S. Agency for International Development (USAID), which provides foreign aid. An avid cyclist, Bruce had read about my trip in *Bicycling* magazine, and had emailed me to offer assistance. Bruce rode a red Cannondale touring bike, and sported a slightly graying mustache and goatee.

"It's really easy to bike into Bogotá," Bruce told me. "There are bike paths all the way to my apartment."

En route we were passed by a number of cyclists on road bikes, wearing Spandex and colorful outfits. Bruce told me cycling is one of Colombia's most popular sports. I replied that more than one Colombian had asked if I knew Lance Armstrong.

In one block the city transitioned abruptly from pastureland to buildings, and Bruce told me that Bogotá has an enforced urban growth boundary. We then went over a busy road by crossing a bridge built specifically for bikes and pedestrians. "These bridges have been built all over the city," Bruce said. "You can see that the bike paths are relatively new." Freshly paved, the two-lane path was about eight feet wide, with a dashed yellow line down the middle. The path turned right into the city, following a grassy lawn that paralleled a river with concrete banks. Though the grass was well manicured, the water was full of brown effluent, reminding me that Colombia is still a poor country.

"How did Bogotá afford these paths?"

"Well, there have been great public infrastructure improvements over the past decade, mostly through the work of two forward-thinking mayors. These bike routes were an interesting story. Each year on their city tax returns, residents can choose to pay an extra tax, choosing what program that extra money should fund. Since enough people opted to pay more taxes for bike routes, these paths were built."

We passed over another bike-pedestrian bridge, a modern-looking suspension bridge slightly narrower than the average car speeding on the busy road below. At the top of the bridge I gained a view across the city. Few of the buildings climbed higher than ten stories, and most were made of brick. A line of green mountains rose and formed the city's eastern border, and below us a five-lane road ran in each direction. The inside lane was nearly empty and separated from the rest of the traffic by a yellow curb; thus unobstructed by traffic, a red double-length articulated bus drove down the lane.

"That's the TransMilenio bus service," Bruce remarked. He explained that it's more or less a metro system—built in just the past few years—with above-ground buses instead of a subway. The buses have their own lanes and never have to wait in traffic. "It's cheap and very successful."

"What's it like?"

"I *hear* that it's great. As an employee of the U.S. government, I'm not allowed to take public transit in the city for security reasons." He calmly explained that kidnappings had only decreased in the past few years.

We rode to Bruce's apartment building, a tall brick structure in a wealthier section of town. I planned to split my time in Bogotá between Bruce's apartment and that of Ricardo, the bicycle advocate who'd originally convinced me to visit Bogotá.

The following day I met up with Ricardo. He wore thin rimless glasses on his angular, professorial face, and his curly dark hair tapered to short sideburns. Biking together across a section of Bogotá, we pedaled down the bike path beside the road. The wide sidewalk had two painted lanes: one for bikes, and one for pedestrians. But since these were not well separated, we rode slowly to avoid hitting the numerous pedestrians. As the bike path dipped down a ramp to cross a street, Ricardo leaned over to me.

"I know this city like no one else knows it. I've crossed it every possible way by bicycle. I know the details."

I had met Ricardo in Mexico City, many months before. He had convinced me to come to Bogotá when he told me about its weekly Sunday morning *ciclovías,* when the major roads are open only to bicycles and hundreds of thousands of Colombians ride the streets. Also, the city boasted over 180 miles of paved bike paths. Ricardo runs the organization Ciudad Humana (Human City), a Bogotá nonprofit dedicated to making cities more livable. The organization focuses largely on transportation issues and, of course, promotes bicycle use.

Ricardo and I veered off the bike path and followed a single-lane road where we jostled for position with small yellow taxis. Ricardo rode upright on his commuter bike, which he'd purchased while studying for his urbanism PhD in Paris. Ricardo was fluent in French but, while he read English, he barely spoke it. I was a bit surprised to meet a highly educated person who didn't speak English; it made me realize how spoiled we English speakers are. "The bike paths don't connect," I yelled to Ricardo in Spanish, in frustration over the noise of traffic.

"It's an incomplete network. Only half done." We continued, navigating potholes until we encountered another main road that, fortunately, had a parallel bike path.

The broad sidewalk had concrete bollards, spaced so that no car could pass. Ricardo noted, "You see all of these bollards? They were built by Mayor Peñalosa's administration. This sidewalk used to be roadside parking, but Peñalosa decided to *decrease* parking, and make more spaces for pedestrians and bikes."

"How did people respond to that?"

"They didn't like it at first!"

I followed Ricardo to the National University of Colombia campus, where he sat on a panel discussion about transportation in Bogotá. Following the talk, I joined him for a meal at an Italian restaurant. I asked again about the restrictions on car use in the city. "Well, Peñalosa, who governed from 1998 to 2000, basically declared a war against cars. Peñalosa said 'private automobiles are the worst threat to quality of life in this city.' In the mid-nineties, most citizens claimed that congestion was the biggest problem in Bogotá, bigger even than crime or the kidnappings. It would take three hours to cross the city by car at the peak hour. Peñalosa restricted driving through a number of different methods." Ricardo explained that one method was limiting the weekdays any particular car could be on the road. Although people initially resisted, because of these measures and better public transit, traffic has decreased by 40 percent. Moreover, because of the TransMilenio, it now takes less than an hour to cross the city. Far more successful than expected, it now provides 1.5 million trips per day in the city. Bogotá has also built bike routes across the city, which has doubled trips taken by bike.

"I get the impression that Bogotá is a completely changed city," I said.

"Yes, and it isn't just infrastructure." He explained that even the crime rate is far down. The murder rate is four times lower than it was fifteen years ago. Bogotá now has a lower murder rate than Washington, DC.

The waiter briefly interrupted us, setting down two artfully arranged plates of pasta. As I dug into my spaghetti, Ricardo told me about other transformations in the city. A different mayor, Antanas Mockus, implemented a public education campaign to improve traffic manners. This campaign included advertisements requesting that citizens cut back on honking their horns; he also hired mimes who, positioned at intersections, mocked anyone disobeying traffic laws. The infrastructure improvements and citizen-awareness campaigns have created a stronger sense of unity. "People are proud of Bogotá now. That wasn't the case a decade ago."

Ricardo and I then discussed why more people don't bike, and the challenge of making bicycle transportation more appealing. While cycling in Bogotá had more than doubled during the previous decade, it still accounted for only 4 percent of trips taken in the city—whereas, in European cities like Copenhagen and Amsterdam, over 30 percent of daily trips are by bicycle.

"I know people would be healthier and probably happier," I said, "if more biked for transportation. Yet I know not everyone has that option, and I don't feel comfortable telling people they should bike instead of drive. I feel like I'm scolding them, and I don't like that."

Bogotá's *TransMilenio* bus

"Well," said Ricardo, "if we design our communities to be more conducive to bicycling, we hope people will make the decision for themselves. Here in Bogotá, personal vehicle use has decreased, bicycle use and walking have increased, and people like the city more." Ricardo paused, and added, "Did you know we have a Day Without a Car here in Bogotá? Once a year, on a workday, we have an event called Day Without a Car, where no one uses a personal vehicle. Everyone walks, uses taxis, or takes the bus. It's a wonderful day—the air is clear, the streets are more pleasant. It shows what is possible in a city."

Something else remarkable had happened in Colombia in the decade before my visit: the country had reduced its carbon dioxide pollution. Some of that reduction derived from increased hydroelectric power—80 percent of the country's electricity comes from dams—and decreased coal-fired power. But the TransMilenio and bikeways also had a serious effect, perhaps decreasing Bogotá's pollution by over half a million tons of carbon dioxide a year, and cutting Bogotá's total pollution by a few percentage points. Car use in Bogotá had dropped significantly, with nearly 20 percent of daily trips now taken via the TransMilenio, an efficient service that didn't even exist a decade earlier. Bogotá has demonstrated that reducing pollution often has ancillary benefits. In fact, the city didn't set out to reduce pollution; the city set out to make itself more livable—and just happened to reduce its fossil fuel use in the process.

If cities in the developing world decided to copy Bogotá, it could make a huge difference for global pollution. In the next thirty years, almost all growth in greenhouse gas pollution is expected to come from developing nations such as Colombia—nations where living standards are rising rapidly. With cities in these countries growing quickly, decisions made today will determine the transportation infrastructure—and contribution to climate change—for decades to come.

Of course, many cities making infrastructure decisions look to other cities for models. To compare two extremes, let's consider Los Angeles, where the majority of commuters drive, and Copenhagen, where transit is evenly divided between personal automobiles, public transportation, and bicycles. Comparing these two cities' annual production of carbon dioxide from transportation: the average citizen of Los Angeles produces about five tons, whereas the average citizen of Copenhagen is responsible for less than one and a half tons.

As over half of the world's population now lives in urban areas, the difference between this number of people living in Los Angeles versus living in Copenhagen would be twelve billion tons of carbon dioxide per year. In comparison, in 2009 all of the world's transportation was responsible for less than seven billion tons of carbon dioxide. In other words, the difference between a world of Los Angeleses and Copenhagens is enormous.

In terms of managing its pollution, Bogotá appears to have moved closer to Copenhagen than to Los Angeles, and I hope more developing cities will follow suit. I also believe that a city that fosters public transit and bicycling is a much more pleasant place to live in. Of course, my bias is strong—I've spent far too much time on a bike. But, at least for now, it would seem that the people of Bogotá would agree with me.

The following day, before work, Ricardo and two employees at Ciudad Humana joined me to explore the city by bicycle. Wearing long pants and light jackets, we departed before sunrise, when the Bogotá air was cool but not cold. As we pedaled to the southern end of town, passing two-story brick buildings cramped together along the road, I noticed that the main roads were dirt, while the bike path was paved. I soon saw why: as dawn approached, bike traffic swelled. Most bicycles were simple one-speed bikes, and most riders were men, though we did see the occasional woman, often pedaling slowly with a small child perched on the top tube. We stopped at a TransMilenio bus stop, where I saw firsthand how efficient Bogotá's system is. Rather than stopping at every block, like many buses do, the TransMilenio bus stopped at only every few blocks; each bus's stopping time was also shortened by the fact that passengers bought their tickets before boarding. And for those who wanted to bike *and* take the bus, Ricardo showed us the bus station's guarded bicycle parking. Cyclists can bike to the TransMilenio, park their bike, and then take public transit across the city. "This bike parking service was my idea," Ricardo said with a smile.

Morning commute in Bogotá

DAY 227
June 19—Chingaza National Park

During my last day in Bogotá, Andres, a young geography professor at Universidad de Los Andes, gave me a ride in his small Renault into the mountains forty-five minutes northeast of Bogotá. Having learned about my journey through a friend in the States, Andres offered me a tour of the nearby Chingaza National Park. We followed a road up through fields of potatoes, continuing to climb after pavement turned to dirt. Along the way Andres described studying for his PhD at the University of Florida. In perfect English he said, "Americans never knew where Colombia is on a map, and even if they did, they'd always ask if I drove home. Haven't they heard of the Darién Gap?" I sheepishly admitted that I had never heard of the Darién until I planned my trip.

When we passed through the gate to Chingaza National Park, cultivated land gave way to short, spongy grass. We stepped out of the car into the cool, thick clouds. Having planned ahead, I was wearing all my warm clothes and rain gear. "Welcome to the *páramo* ecosystem," Andres said. We were at about 13,000 feet; though we were in the tropics, the temperature hovered in the mid-40s.

Moisture condensed on my rain jacket as a stiff wind blew against us. Following a trail up the hillside we passed grasses, a variety of small shrubs, and mosses clinging to rocks. A small stream of water flowed down the trail, which I ascended in socks and sandals—still my only shoes—with plastic bags over my socks to keep warm. With each step, my feet sunk into the ground as if walking on a wet mattress. As we climbed, the landscape became dominated by what looked like tiny palm trees, ranging in height from one to ten feet and covered in dew. A few had blooms resembling sunflowers. "Those are *frailejones*," Andres explained, raising his voice over the wind. "As it's always cold here, they grow slowly. Some of the big ones are over a hundred years old."

"You said before that this region is highly susceptible to climate change," I half yelled back to him.

"Yes," Andres answered. "Human-caused climate change will create a climate far outside the realm of natural variation. This area was glaciated during the past ice age." The signs of the last ice age were everywhere around us—lakes in ice-carved bowls in stone, and large boulders called erratics, left behind by glaciers. Andres explained how, during the Earth's various ice ages, the *páramo* ecosystem could be found at a lower elevation—one could have found these plants in Bogotá's valley. But now the ecosystem has retreated to the mountaintops. If climate change warms this area, the ecosystem will have nowhere higher to climb. Andres told me that a recent study estimated that if the Earth warms by 2.5 to 3 degrees Celsius (4.5 to 5.5 degrees Fahrenheit), we'd lose 97 percent of the *páramo*.

"You said this place was important for Bogotá's water supply, right?"

"Yes, and the water is very clean. You can drink it. Watch." Andres scooped up water from a small stream and drank it. I filled my water bottle and took a sip; it was cold, clear, and pure. Andres then explained how the spongy ground acts as a huge reservoir, which slowly releases water in a constant stream. Since the ecosystems at lower elevations have less organic matter in their soils, they can't store as much water. The bulk of Bogotá's potable water comes from the national park lands.

Andres said, "You know that Colombia has perhaps the greatest biodiversity of any country in the world? We have the *páramo*, we have the Amazon rainforest in the southeast, and we even have a desert in the far northeast. The *páramo* alone is home to some two thousand species of plants and animals, many of which are found only in Colombia. All that—all of this—is at risk."

Páramo ecosystem

"Aren't the mountains taller in Peru and Bolivia?" I asked. "Why couldn't these species survive there?"

"It is drier down there—a different ecosystem. These plants are found only in Colombia, Venezuela, and northern Ecuador."

The clouds occasionally lifted, revealing lakes and the rolling hills covered by grass and the *frailejones*. Tufts of cloud moved with the wind and combed the land. As I had already done several times along my journey, I again found myself in a beautiful, diverse ecosystem that might not still be there should I return.

DAY 228
June 20—Pamplona

Before leaving Bogotá, I interviewed with *El Tiempo*, Colombia's largest newspaper, which printed and posted online an article about Ride for Climate. Soon after, I received email invitations from two different Colombians living in the States. The first, Diego, studying at the

University of Nebraska, invited me to stay with family members living in Pamplona and Cúcuta; the second, Mauricio in California, offered lodgings with his family in Bucaramanga.

So I departed Bogotá and continued northeast toward Venezuela, following the Andes north on a route free of guerilla or paramilitary forces. I climbed and descended through numerous valleys, encountering deserts at the bottom of canyons, broadleaf forests at mid altitudes, and *páramo* at the tops of the climbs. High in the mountains I wore my BOMBEROS DE MEDELLÍN jacket, and more than once was asked if I fought fires in Medellín.

When I reached Diego's aunt in Pamplona, she asked how I knew him. I surprised her with my reply: "I've never met him. He read about me on the Internet." Nonetheless, she welcomed me into her home, feeding me potato soup and fried trout from a nearby mountain lake.

Warmed by the generous hospitality of strangers, I had a moment to catch my breath and think back on the past month. Though at first exhausted by the heat and frustrated by my "off route" ride to Bogotá, I had fallen in love with Colombia. The country offered cloud-covered mountains, deep-green hillsides, bike lanes, and *ciclovías*—and everywhere I went I encountered citizens eager to paint a different picture of Colombia than is portrayed in the news.

I was enjoying feeling so far away from home, enjoying being lost in my journey. I also realized that one reason I felt so far away from California was that, unlike those I'd met in Mexico and Central America, few of the people I met in the Colombian countryside had family in the United States. (In preventing car traffic between Colombia and Panama, the Darién Gap makes it difficult for South Americans to venture north.) I was now on a different continent than when I started, and it felt that way.

I'd also fallen in love with Colombia's bike culture. I still smile when I think of how I'd arrived in Medellín on a Sunday morning to find the major roads open only to bicycles, and how I biked across Bogotá following bike lanes safely separated from car traffic. I find cities that are safe to bike in intrinsically more welcoming. Cars can be noisy and dangerous, and they wall us off from our neighbors. Bikes, on the other hand, create no boundaries between individuals. As such, they are a statement of trust and freedom. They can also be a part of the solution to climate change. Bike culture is not the entire solution, as no single remedy will solve this challenge. But I firmly believe one excellent path to cutting carbon emissions and improving our quality of life concerns two interrelated elements: better urban planning, and the bicycle.

COLOMBIA
May 25–July 4
Presentations: 3
Flat tires: 6
Miles: 1,123
Trip odometer: 7,232 miles

Venezuela route

14 BOLIVARIAN REPUBLIC
OF VENEZUELA 🚲
Petro-Democracy

Venezuela population: 30 million
Annual per capita GDP: $13,500
Annual per capita CO_2 emissions from fossil fuels: 6.9 tons

DAY 243
July 5—Cúcuta to San Cristóbal

How can there be so many gas-guzzlers? I asked myself after crossing the border and seeing the line of cars waiting to enter Colombia from Venezuela. There were a surprising number of new SUVs, as well as used Chevys and Fords from the 1970s—all boxy cars with giant hoods and trunks. As the cars inched forward toward Colombia, I saw that the line of vehicles disappeared into the mountain road ahead like the tail of a long snake.

As I rode the opposite direction, in a mostly empty lane leading into Venezuela, a warm but fierce headwind blasted against my face, as if warning me against entering the country. Weary from an hour spent battling the wind and riding uphill, I turned into a pullout to rest. I leaned *del Fuego* against the metal railing just off the road and took a seat next to a roadside shrine marking the site of an accident. I felt weak and tired. Having come down with a severe cold, I had spent most of the previous week in a small town in Colombia. When not sleeping or resting in a hostel, I'd spent my time surfing at the small Internet café, watching movies at a restaurant, or talking to the town's numerous M16-toting police officers, all of whom were eager to say "Hello Meester," in English. The cold had delayed me almost a week, and had sapped the energy from my legs.

Now in Venezuela, regaining my strength at the rest stop, I startled when a large gray box-like car veered out of the line and drove straight toward me, screeching to a stop. Out from the vehicle burst a balding, slightly pudgy man, followed by a woman and a boy who was maybe eleven years old.

"We saw you on television today!" The man spoke with excitement as the woman smiled brightly. Earlier in the day I'd been interviewed for a live morning show in Cúcuta, the Colombian town just across the border. The man then said, "Here, this is to help you on the road"—and he handed me two blueish purple bills, each worth five thousand bolivars. Having just learned yet another exchange rate, I made the conversion in my mind. *Ten thousand bolivars is about . . . five dollars.*

"Wow, thank you, this is very kind," I told them. "Where are you going?"

"We drive across the border every day, fill our tank with Venezuelan gasoline, return to Colombia, and siphon off the gas to sell."

"You can make money doing that?"

"Yes, gas in Venezuela costs seventy-five bolivars per liter." I converted from bolivars per liter to dollars a gallon . . . meaning that gas cost . . . *fifteen cents a gallon?*

"In Colombia," I said, "it costs twenty times that amount."

"Yes," the man replied, "it's a great deal in Venezuela. They give it away."

"It can't be legal to do what you're doing, *is it?*"

"Of course not, but everyone does. What do you think all these cars are doing?"

The family wished me good luck, returned to their car, and drove back into the line of vehicles waiting to enter Colombia.

———————

Venezuela is blessed with huge oil deposits. The world's fifth-largest oil exporter, it's the United States' third-largest supplier of oil after Canada and Saudi Arabia. Not surprisingly, the petroleum industry accounts for over 85 percent of Venezuela's exports, and provides the government with half of its revenue.

When I designed my route, I planned this large detour east through northern South America partially because I wanted to see a country where oil dominates the economy. I know oil contributes heavily to greenhouse gas pollution, but what about its social and economic implications? Many people have argued that oil wealth funds terrorist organizations in the Middle East, and some claimed that Hugo Chávez, Venezuela's leftist president at the time, similarly provided financial assistance to the FARC in Colombia.

I had also heard of the "curse of oil"—the counterintuitive claim that vast oil resources actually do a country more harm than good. The curse theory claims that the massive flow of cash to government coffers hinders democracy by allowing corrupt regimes to function without accountability, and that oil distorts the economy and slows development. This curse is hard to believe—one would think the more oil a country has, the better off it would be. Yet many of the most oil rich countries are in fact poor, including Venezuela, and many are run by non-democratic autocratic governments. So, puzzled by such discrepancy, I wanted to find out for myself: just how does oil affect Venezuela? To answer that—in an entirely unscientific way—I'd bike across the country and talk to the people I met.

―――――――

After a few more hours' ride climbing into the wind, I descended into San Cristóbal, a city of half a million residents. I passed under a freeway with both on- and off-ramps, noting that no country since Mexico had had such a wide freeway in a small city. Continuing downtown, I found the city's roads had narrow sidewalks, with bumper-to-bumper traffic in both directions. I passed a drive-through restaurant and a drive-through bank, structures I hadn't seen since California. Perhaps I was merely tired, still recovering from my cold, but the traffic depressed me, and made me feel unwelcome on a bicycle.

A large mural on the roadside depicted an oversized close-up of president Hugo Chávez. With a roundish face with short curly black hair,

Chávez wore a red beret and red jacket. His fist was raised, and a crowd of people in red standing behind him held their hands in the air. The sign read Rumbo a 10 Millones, signifying Chávez's goal to get ten million votes in the next election. Farther down the road, another large billboard showed a smiling Hugo Chávez and read Thanks to President Chávez for the Completion of the Uribante Caparo Hydroelectric Dam, as if Chávez had built the dam with his own hands.

Chávez was a fascinating and controversial president. He formed alliances with Iran and Cuba, and took a leadership role in OPEC (Organization of the Petroleum Exporting Countries), of which Venezuela is a founding member. He regularly called George Bush "a pig," and said, "Capitalism leads us straight to hell, and I hereby accuse the North American Empire of being the biggest menace to our planet." Yet the words rang strangely hollow, especially since the Venezuelan government got most of its revenue selling oil to the United States. Chávez also accused the U.S. of supporting a failed coup against him in 2002. (Though there's little evidence that such took place, the Bush administration notably did not condemn the incident at the time.) In turn, Bush accused Chávez of squashing dissent and acting non-democratically, even though Chávez won three democratic elections—ultimately four. And though Chávez's critics say he used Venezuela's oil money to bully other countries in the region and to buy friends, he also used oil wealth to fund social programs to help the poor, providing health care and education, claiming to be a champion of "Twenty-first Century Socialism."

DAY 244
July 6—San Cristóbal to Lake Maracaibo

After my first night in Venezuela, which I spent at the San Cristóbal fire station, I looked over the map and concluded I wouldn't make it to Caracas, Venezuela's capital, soon enough. A good friend was flying to meet me there, but my week off recovering in the mountains had put me far behind schedule. I had already realized that maintaining a strict schedule while traveling by bicycle is a real challenge, as illness or mechanical failure was always just around the corner. So I had been padding my schedule to allow for rest days and possible delays. If in the end no troubles arose, I could always spend the extra time visiting schools, exploring cities, or updating the website. But if I was a few days behind where I planned to be, I could ignore everything and focus on biking to catch up.

But even with that planning, I was too far behind to make it to Caracas on time. Also, I wanted to visit the oil fields of Lake Maracaibo, which was far off my route. I didn't feel bad about skipping part of my ride across Venezuela, but didn't know how should I travel—by plane? By bus? Either would be cumbersome with *del Fuego*.

Cheap gasoline, though, provided a solution to my problem: the country had many large cars and trucks, including numerous pickups with empty beds. Perhaps I could hitchhike a ride for two on one of these trucks.

So, from the San Cristóbal fire station I biked to the freeway leading north out of town. At the first on-ramp I encountered I stuck out my thumb to hitchhike, and after just fifteen minutes, a pickup offered me a ride. I lifted *del Fuego* and its saddlebags into the truck bed and squeezed into the cab, where a man in uniform sat with a woman I assumed was his girlfriend or wife.

"Where are you going?"

"Toward Maracaibo."

"I can take you as far as El Frío."

The miles flew by effortlessly riding in the pickup. Though I had promised myself I would hitchhike only to make it to Caracas, I worried that I'd be tempted to get rides in the future. A car seat is *comfortable*.

"What do you think of president Chávez?" I asked the driver.

"He's a great president."

"Why?"

"No one before him has cared about the poor." He didn't elaborate on his answer.

After descending out of the mountains, the driver dropped me off in a small town. My plan was to hitchhike north to the Caribbean, and then east along the coast until I reached Caracas, which was just inland of where the Andes meet the Caribbean. From there I'd start biking again, traveling east and then south into Brazil. Southern Venezuela is almost pure wilderness, and only one road leads into Brazil, even though the countries share an eight-hundred-mile-long border.

After picking up a few supplies in the town—pasta for dinner, pure alcohol for my stove—I biked north along the two-lane highway looking for a good place to wait for a ride. Two miles down I encountered a roadblock, where army officers stopped cars to check papers. I dismounted and approached.

"Hi, I need help," I told the man. "Here, this is me." I pulled out the *El Tiempo* article about my journey, hoping that my being in a national newspaper would make them more likely to help and less likely to hassle me. "I need a ride—do you think you could ask if I could get a ride north?"

The man wore the Venezuelan uniform—complete with a maroon beret—and didn't smile, but after looking over the article he said, "Sure."

Five minutes later, I was sitting in the bed of a pickup truck next to *del Fuego* feeling the tropical air fly by.

———

After five pickup truck rides and two hundred miles—never waiting more than twenty minutes for a ride and *amazed* at how easy it was to hitchhike with a bike—I departed the main highway and started pedaling down a dirt road toward Lake Maracaibo. This lake, a teardrop-shaped body of water the size of Connecticut, isn't really a lake, as its north end connects to the Caribbean via a four-mile-wide channel, thus making it more of a brackish sea. Beneath the lake are enormous oil deposits. The Venezuelan state oil company, PDVSA, daily pumps more than a million barrels of petroleum from the Maracaibo region.

After biking through mostly uninhabited ranch land, I reached the lake, gaining a view across its calm waters. On the horizon a few oilrigs were barely visible, and along the shore palm trees grew between a few small one-story structures, houses constructed from a mix of wood, corrugated tin, and palm leaves. Walking along the road's edge, a shirtless man with a potbelly and a pockmarked face yelled to me.

"Hi—come join us!" he said, introducing himself as César. He had a big mischievous smile, as if he were up to something but not quite clever enough to pull it off. "Have some beers with us." He lifted a Regional Light, the Venezuelan cheap beer, out of a Styrofoam cooler.

"Umm . . . well . . . we'll see," I said. Though it was nearly dark, and I needed to camp soon anyway, I wasn't sure I could trust this man. I dismounted and walked alongside him. "What do people do in this village?"

"Most work in fishing, but I work on the oilrig."

"Why aren't you working today?"

"I work for seven days on the rig, and then get seven days at home."

César lived in a three-room house with dirt floors that he shared with his wife, son, daughter-in-law, and three of his grandchildren. Some of the walls were made of concrete, others of wooden boards. The roof was a combination of tin and palm leaves; branches held up an extended portion of the roof over the entrance to the house, making a small dirt porch. The bathroom was an outhouse perched on a platform over the edge of the lake. "You can just go in the lake like that?" I asked, pointing to the outhouse.

"The lake washes it away and it's fine."

The lake, which is connected to the ocean and thus is at sea level, was only a foot or two below the level of César's house. These houses were makeshift, built quickly and not expected to last. If sea level were to rise just two feet, all the houses along the beach would have to be rebuilt elsewhere. While new similar shacks may be easy to construct, I wondered where they would be built. Just inland of the structures was privately held grazing land, and someone would have to give up their land. Throughout Latin America, about ten million people might have to move their homes if sea levels rise just three feet.

The family seemed friendly and harmless enough, so I asked if I could pitch my tent and stay the night. César agreed, and offered me a dinner of chicken and rice, which he set on a sort of improvised "TV table"—a square piece of plywood on an upside-down plastic bucket, placed in front of their home. After subtly making sure the food was well cooked, I sat down to eat. As I cut the chicken with my fork, César's family—who had already eaten—and a few of his neighbors stood and stared, watching me chew every bite. Some of the men were shirtless, and everyone wore sandals. Then I realized that my audience extended beyond César's house—I saw a family peering out the window of *another* house, staring with curiosity and surprise. I wondered what they thought of me. My bags were now thoroughly worn, and the brown paint had partially rubbed off my panniers, making them an ugly mix of red and brown. *Del Fuego* was dirty and covered with fraying and faded decals.

César's shower wasn't functioning, so his brother, Hiro, who lived three houses over, offered to let me use his. As I walked over, I asked Hiro what his favorite part of town is. "Shrimp fishing," he said.

César (center back) and his family serve me a meal

"Why?"

"Because that's what I do for a living." Hiro's house, to my surprise, had a concrete floor, and his bathroom had a tile floor, even a porcelain toilet. I was astonished that a fisherman had a nicer home than the man employed by the oil industry, but I didn't feel I should ask about this.

After the welcome shower I returned to César's house. As night fell, the town's small houses lit up, and bright streetlights lined the road. "Electricity is free here," César told me, and then I saw his latest purchases: a small color television, two large speakers, and a DVD player.

César and his son rifled through a small pile of pirated music video DVDs. The new television lay on a tree stump just outside the front door, the speakers sat on upside-down plastic buckets, and the DVD player rested on the ground. "This is Colombian music," César said, slightly slurring as he drank another beer. "I'm from Colombia, but moved here."

Though César explained he was from Caucasia, he showed little interest when I told him I'd biked through the town and visited the local school. As I'd found with many people I stayed with in the countryside, he asked few questions beyond the value of my bike. I wondered if perhaps people thought the cultural divide between us was too great, that they just didn't know what questions to ask.

Soon, music videos blasted from the television, and César drank more beers with his son. The videos featured Colombian men playing accordions and singing while bikini-clad women danced around them. César's wife and a few other family members sat watching quietly in plastic chairs, while a friend of César's son lay in a hammock, also drinking. César's daughter-in-law, who appeared to be in her late teens, stood in the doorway with a newborn in her arms, looking somewhat uncomfortable as the men drank beer and sang along to the videos.

I turned to César, "Don't you think you should turn the music down? It's past eleven. Aren't most people in this town fishermen who'll get up early tomorrow?"

"Don't worry about it," he replied, more heavily slurring his words.

I retreated to my tent along the banks of the lake, and lay on top of my sleeping bag in the hot breezeless night. I was depressed about César's treatment of his neighbors, and maybe about the confusion I felt over the discrepancy between his situation and the clearly better one of his brother—not to mention that César spent his money on entertainment components instead of installing a sanitary toilet. I also felt a bit depressed about judging him this way, especially given his hospitality.

DAY 247
July 9—Ciudad Ojeda

After biking to the city of Maracaibo and spending a night at the fire station, I hitched a ride around the northern edge of the lake and across the five-and-a-half-mile bridge that stretches over the waterway connecting the lake to the ocean. I then biked along the eastern shore, following a path at the top of a levee separating the water from settlements. Venezuela's largest oil deposits were found along this shoreline. Over the past century so much oil has been extracted that the land has sunk by many feet—and in some places as much as twenty feet. The cities along the lake's northeast coast are now slightly below sea level, protected by the long levee.

In the lake small, evenly spaced oil derricks spread to the horizon. Some protruded up like towers, while others bobbed up and down.

Covering about half the surface of the water was a green film, duck-weed, a floating plant that had taken over the lake, making it look like an enormous waterbed covered by a green sheet.

I looked to my left, where the land was oddly at a lower elevation than the lake's surface. From the ground pipes led up, over, or through the levee into the water. After every rainstorm water had to be pumped uphill to sea level, as otherwise water would pool in the town. *Is this a vision of the future?* The sea will rise—and either people will migrate away or these levees and pumps will be built along every shoreline.

Oil derricks on Lake Maracaibo—duckweed covers the lake's surface

Moving on from the levee I rode into Cabimas, a shoreline city originally built by the Shell Oil Company. I passed a large fenced-off facility with oil towers behind it. When I asked a security guard if there were tours of the oil wells he laughed at me. "No. Of course not. You're not even allowed to take pictures of this facility!" I biked on, occasionally encountering small pump jacks bobbing up and down between wooden houses. The downtown was choked with car traffic, and cars crowded in front of the small white cathedral at the Plaza Bolívar. I kept thinking about the contrast with Colombian towns—which were mostly free of traffic—I had crossed two weeks earlier.

As I turned down one of the secondary roads leaving town, a woman stopped me. "Don't go down that road," she said, "it's not safe."

I soon realized what she meant: I felt uncomfortable—the eyes of the locals were fearful, distrusting. I thanked her for the warning.

I stayed on the main road farther inland and rode toward the next town, Ciudad Ojeda. A few hours later, as darkness approached, a large SUV with shiny black paint pulled up alongside me, slowing to the speed I was biking. When the dark-tinted window rolled down I saw a blond woman in the passenger's seat and a skinny man with glasses driving. He leaned over while holding the steering wheel and yelled in English, "Do you need a place to stay? I saw you in the paper today."

"Ahh . . . sure . . . ," I said, without stopping.

"You can throw your bike in the back of my car," he offered.

"Can I just bike along and follow you? It's a pain to take the bike apart." I wasn't yet ready to commit myself and my belongings to this stranger.

"Sure. We live close." The man introduced himself as Juan and the woman as Victoria.

The couple led me about a mile to a high-rise dubbed New York Place, a modern building where armed security guards protected the parking lot and main entrance. The ten-story building rose up next to makeshift shacks only slightly nicer than César's house. Seeing I'd be in a protected area, I decided to accept their offer. I joined them in an elevator to the eighth floor and rolled *del Fuego* into their apartment. It was bright and clean, with a finished wood floor, new comfortable furniture, a high-quality sound system, and a projection TV—an apartment a thousand times more comfortable than César's shack on the lake. I felt incredibly out of place—I was covered in sweat, and *del Fuego* at my side was filthy. Fortunately, Juan and Victoria didn't seem to mind. When they showed me their guest room, they admitted the room had never been used. We ordered dinner from Pizza Hut. "We don't usually have time to cook," Juan said. I noticed the kitchen was spotless.

I learned that Juan was from Colombia and Victoria, his girlfriend, was from Argentina. They both worked as engineers for Schlumberger, a European company contracted to drill for oil. Juan was the manager of the local office.

"Where else have you worked?" I asked.

"All over the place," Juan said. "Argentina, Angola, Houston, and now here."

After briefly lamenting that there was little oil in his home country of Colombia, Juan explained what it was like working in Houston.

He loved the United States. "You can buy or do anything there," he said, then told me he'd even bought a small plane and learned how to fly. But he did complain that it was extremely difficult to bike to work in Houston; it turns out Juan was also an avid cyclist.

"You tried to ride your bike to your oil job?" I laughed. "Doesn't that seem like going against the company?"

"Well, I gave up quickly. It was impossible to ride in Houston. Too many cars."

Over a sausage-and-ham pizza and a two-liter bottle of Coke, I asked Juan and Victoria about what it's like to drill for oil.

"Well, we take a boat out to a rig and then drill about three thousand feet under the floor of the lake," said Juan. "Victoria and I have to analyze the rock as we drill, and make quick decisions about whether to start pumping, keep drilling, or go somewhere else. Drilling can take thirty-six hours, and we can't take a break because the hole might collapse, so we have to stay awake and focus for a very long time. It's hard work and takes a lot of training."

"Ahhh . . . you know," I said, pausing. ". . . you read about me in the paper. I'm biking to draw attention to global warming." I wasn't sure what Juan and Victoria would think of my project or climate change. I'd never actually discussed the subject with a petroleum engineer.

"Yes," replied Juan, "you know, oil companies get a bad rap, but the truth is that people need oil. We drill because there's demand. Also, oil companies can be part of the solution. We have the technology to pump carbon dioxide back underground—which is one possible solution to global warming."

Pumping carbon dioxide into underground reservoirs just might be part of the solution, though it's also riddled with challenges— it's expensive to remove carbon dioxide from a power plant's effluent to pump it deep into the Earth, and the carbon dioxide may not stay underground permanently. Carbon capture and sequestration, or CCS, may play a vital role in avoiding the build-up of carbon dioxide in the atmosphere, as it would allow us to continue to burn some coal, gas, and oil as we develop cheap renewable energy. Since oil companies have the expertise to drill into rock and pump the carbon dioxide underground, we will need their help to make CCS a significant part of the solution. This is true especially given the daunting numbers: if we were to use CCS on all coal power plants in the world, we would have to annually pump around twenty billion tons of carbon dioxide underground.

After another bite of pizza I asked, "What do you think of the future of oil use?"

"Well, there's no fuel as useful," Juan said. "Per gallon, gasoline has more energy than almost any other energy source. People are going to be using oil a hundred years from now."

"You don't think we'll run out?"

"Well, all the cheap and easy-to-get oil is gone. The oil we drill for now is difficult to extract. We're drilling three thousand feet deep here. When I worked in Houston, we drilled ten thousand feet deep—it's so hot at that depth that our drill bits melt. But there's still a lot of oil out there. In eastern Venezuela we have about a trillion barrels of oil underground."

"Ah . . . how much is that?"

"About twice as much oil as in the Middle East."

"*What?* How can there possibly be more oil here than in the Middle East?"

"Well, the oil in the Middle East is much cheaper to extract and refine—so-called 'conventional oil.' But there's a lot of non-conventional oil in the world—you may have heard of heavy crude oil or oil sands. Canada has another trillion or so barrels in oil sands. This oil is difficult to extract or refine, and currently only a fraction of non-conventional oil is economically viable at current oil prices. But if the price of oil went up enough to make the effort worthwhile, or if we get better technology to extract all of this heavy crude, there's no reason we can't keep using oil well into the twenty-second century."

"I'll be honest," I said, "I'm not a big fan of oil, although you're right, we do need the stuff. Here's another question: Is it good for a country to discover oil? I know it can bring money to a region, but if that's so, why is Venezuela so poor?"

Victoria, who had been mostly silent, said flatly, "No, it's bad." She then described her time in Angola, a country where two-thirds of the population doesn't have access to clean water. "I've seen small villages where the people give up farming once the oil is discovered, and years later when the oil dries up, they don't know how to go back to farming, and they're poorer than before the boom."

While Juan clearly differed with Victoria to some extent, he surprisingly agreed with her on this point. "Look here at Ciudad Ojeda," he said. "This place is a dump, nothing more than an oil camp. We *don't do anything in town*. We don't even shop for groceries here. There are no universities nearby, no public infrastructure."

I was shocked by the simplicity and honesty of this reply—I expected at least nuanced evasion. They implicitly conceded what I thought: the oil industry has little incentive to invest in the towns, cities, or countries where they find oil.

When the conversation turned to politics, I asked Juan to explain how Hugo Chávez came to power.

"You have to understand the history," he said. "Over the past forty years, while many other Latin American countries had dictatorships or violent civil wars, Venezuela had a peaceful democracy, with the presidency alternating between two different political parties. But both parties were corrupt and did little to help the poor, who make up most of the country. In 1992, Chávez, then an army captain, led a coup to overthrow the president. The coup failed, but it gained Chávez national attention and popularity. That should tell you something about the popularity of the old government! By attempting a coup against the president, Chávez gained popularity! Chávez went to jail, but was pardoned two years later by the next president. In 1998, Chávez was elected president, and called a vote to rewrite the constitution."

Juan continued. "It was quite brilliant. The constitution was rewritten to give the president far more power than before. Elections for every office in the nation took place, and because Chávez was popular at the time, all levels of government were replaced with his supporters."

"What do you think of the social programs he's initiated?" I asked.

"Well, in the past few years oil prices have jumped, meaning the government has received incredible amounts of cash, and Chávez spends it however he likes. He uses it to bribe foreign governments and fund various new government programs. No one had paid attention to the poor before, so he gets huge political gain for doing so." Juan then explained the various social programs, known as "missions," Chávez established to help the poor: Mission Robinson works to eliminate illiteracy, Mission Barrio Adentro provides healthcare for the poor, and Mission Mercal helps provide discounted food. But Juan described the programs as if they were merely political buyouts, that they didn't actually benefit the country. "If oil prices drop, we're in serious trouble," he said. He described how he thought PDVSA, the national oil company, wasn't making the right investments, and probably wouldn't meet their extraction goals. Chávez, he argued, was hurting foreign investment, which was needed for PDVSA to recover difficult-to-extract oil. Moreover, in 2002–2003 there was a general strike of workers in PDVSA. Chávez fired and replaced most of them, but the replacements were not as skilled as their predecessors. "We're functioning now only because of high oil prices."

After dinner I retired to the spare bedroom, thankful for both Juan and Victoria's candor and generosity. I hoped that, if Juan was correct in saying we'd still be using oil in a hundred years, that at least he was also correct in predicting oil companies can help solve the problem by pumping carbon dioxide underground. He was right, though, that oil is incredibly useful and will be difficult to replace. We won't be able to replace it solely through conservation, improved efficiency, or better bike lanes. We'll need new energy technology.

DAY 248
July 10—Punto Fijo

In the morning, after updating my website on José's high-speed Internet, I returned to the roadside with thumb extended. It was as if I had a free taxi service to anywhere in Venezuela. I got to the northern coast thanks to two different pickup trucks, after which an obese man named Roberto drove me five hours east in his large van.

Roberto and I chatted during our long ride together. He told me he used the van to drive patients to get free eye surgery. "It's one of the government programs. We give Cuba oil, and, in return, Cuban doctors provide free eye surgery. I drove a number of patients to the airport to fly to Cuba— the flight is also paid for by the Venezuelan government."

"That's amazing. Are all these missions helping the country?"

"Well, some people can see now, but mostly the programs don't help," Roberto said matter-of-factly. "Chávez is mostly talk." I found his response odd, given that the government program paid his salary.

When we stopped for gas, Roberto filled the tank for about three dollars. For the last leg of the trip I fell asleep in the passenger's seat, sleeping until Robert dropped me off in the town of Santa Ana de Coro at 9 PM. I biked to the town's fire station. After showing off my collection of firefighter patches, I was given an air-conditioned room for the evening.

The hitchhiking had given me extra time to explore, so the following morning I rode out on a long peninsula that jutted away from the north coast, heading to Punto Fijo and the site of Venezuela's largest oil refineries. Crude oil from Lake Maracaibo was shipped to Punto Fijo, where it was refined before being transported to foreign shores.

After riding fifty miles through desert terrain, I arrived at Punto Fijo. I first noticed that a series of one-story buildings had bars over most windows. I rode toward tall smoke stacks on the town's edge, where I encountered a fence enclosing the mile-wide refinery. Beyond the fence, pipes, smoking towers, and large circular five-story vats occupied an un-human landscape. A stiff, dry wind blew the smoke horizontally from the refineries' tops, the white plume quickly dissipating. I found the twisting pipes both appalling and beautiful. The refinery itself, though, wasn't welcoming. Trash lined the edge of the fence, and signs forbade both trespassing and photography. I biked along the fence until I arrived at an entrance where armed guards checked the IDs of drivers entering the complex.

An oil refinery heats crude oil and separates it into its numerous chemical components: gasoline, diesel fuel, asphalt base, heating oil, kerosene, propane, lubricating oils, paraffin wax, and other materials, all of which are merely strings of carbon and hydrogen. The different components can be separated in different ways, but the process usually requires numerous chemicals, as well as heating the crude to high temperatures. These various liquids and gases don't just provide our fuel—they provide almost every synthetic material we use.

Staring at the facility, I marveled at how almost everything we do in modern life requires these materials. I strained to think of a product or activity that doesn't use oil, grease, or plastic. Just looking down at *del Fuego* and myself was telling enough. The tires, made of synthetic rubber, began their life as chemicals processed from crude oil. The grease in the ball bearings was a petroleum product, as was the plastic components of the shifters. My nylon and plastic panniers derived from petroleum. All these chemicals were once heated to hundreds of degrees and processed in a refinery.

Refinery in Punto Fijo

And what about my clothes—my Spandex, my helmet, my sandals? The materials to make all of these came from a refinery. And while as a cyclist I don't need to buy fuel for a car, the food I eat to fuel myself relies on petroleum to power the tractors that work the fields, and then the trucks that convey the food to stores. And when I reached Tierra del Fuego, petroleum would fuel the plane flying me back to the States, just as it had transported me to vacation in Florida, and transported my friends and father to join me on my journey. Oil may ravage the environment, but the benefits it brings humanity are enormous.

As I was unlikely to receive a tour of the refinery's facilities, I cycled a loop around the perimeter fence and then returned to the highway to hitch a ride to Caracas. I was quickly picked up by José, a man driving a forty-year-old pickup. José told me he worked in Punto Fijo as a carpenter, so I asked him what effect the refinery has had on his town.

"It's bad," José said over the roar of the truck's old engine. "Before the oil refinery, people were poor, and they had nothing to do, so they would drink, smoke, and womanize. Now they have more money—so they can drink more, smoke more, and womanize even more. You have to educate people and give them culture, you can't just give them money." When I asked if he'd heard of global warming, he said, "Yeah, I bet someday you won't even be able to go outside. But you know what I fear more than anything else? The police. They're so corrupt."

At 11 PM, thanks to hitching two more rides, I was dropped off at a tollbooth about a hundred miles short of Caracas. I decided to find a place to camp instead of getting a hotel. Hotels were expensive in Venezuela; in fact, everything except gasoline was expensive to travelers. A hotel room would cost twenty dollars, as compared to five dollars in Colombia. A lunch of *arepas,* a fried corn patty, was six dollars instead of two or three. A small bottle of the pure alcohol I used to fuel my stove—five dollars instead of one.

The reason? Oil. Oil distorts the exchange rate. Since Venezuela sells oil overseas, the value of its bolivar increases relative to other currencies. That's because the purchase of Venezuelan products by non-Venezuelans essentially requires the conversion of other currencies into bolivars—in effect, one "buys" bolivars with U.S. dollars or other currencies, and then buys Venezuelan products with those bolivars. This means that, as the world wants more Venezuelan oil, the "value" of a bolivar increases relative to other currencies. Thus, I could buy less with one U.S. dollar in Venezuela than I could in neighboring countries.

In theory, this suggests Venezuelans get a great rate buying imported goods. The problem, though, is that many Venezuelans don't have jobs. The currency distortion also hurts local industry—increased purchase of imported goods decreases the demand for local goods, so local industries suffer. So, essentially, the non-oil segments of the economy are crowded out by the oil-extraction industry. But as oil extraction doesn't require many workers, it does not employ a large percentage of the population; as a result, an already poor distribution of wealth gets skewed even further.

This distortion of the economy is one reason people speak of the "curse of oil." Indeed, during the last three decades of the twentieth century, OPEC countries' economies grew roughly half as quickly as those of non-oil producing nations. Despite—or perhaps because of—Venezuela's status as the world's fifth-largest oil exporter, average Venezuelans were poorer during my trip than they'd been three decades earlier.

Using oil money to help the poor seems like a good idea because it offsets the tendency of petroleum to increase inequity. Yet the few oil-rich countries that have escaped the so-called "curse," such as Dubai, Mexico, or Indonesia, have done so not by increasing government spending, but by diversifying their economy away from oil and investing in manufacturing or service industries.

Looking for a place to spend the night, I stopped at a parking lot and befriended the security guard, who let me camp in the lot. While I set up my tent, I asked the guard, "Do you think oil is good for Venezuela?"

"It didn't used to be!" he said enthusiastically, "It never helped us before. But now Chávez is doing something with the oil. He has the Bolivarian Dream!" Simón Bolívar, considered South America's George Washington, liberated Venezuela, Colombia, and Ecuador from the Spanish in the 1810s and 1820s. Chávez invoked Simón Bolívar's name as much as possible, and he even officially renamed Venezuela to the unmanageable "Bolivarian Republic of Venezuela." Among other claims, Chávez said he was attempting to achieve the equality that Bolívar had once envisioned.

"How is oil good for Venezuela?" I asked the guard.

"Well, for one, I have health care now. Before Chávez we were ignored."

"But wasn't there health care before?"

"We have more doctors now."

"Have you personally received better care?"

"Well, no, but I haven't needed it." In theory, the poor now had better access to health care, although I had heard criticism that few actually receive service. But, at the very least, people like this security guard certainly had more pride in their country than they had before Chávez took office.

DAY 250
July 12—Caracas

The following morning I caught a lift to Caracas, riding with a man who said nothing to me the entire ride. The air blowing through the truck's open windows was warm but not hot. After an hour and a half, the two-lane highway dropped us into the long valley of Caracas, from which steep green mountains rose up like a wall behind a string of buildings. From a distance, the city appeared like a slightly warmer and more populated version of Colombia's Medellín.

But closer up the similarity faded. The buildings in the city's center looked as if they'd been built twenty or more years ago. A thirty-story concrete high-rise read NESCAFÉ in red letters across its top; on its roof stood an eight-story-tall red coffee cup, as if waiting to serve a wandering giant. Two buildings over, a similar building read PEPSI across its top floor, and on its roof sat a large globe with the red, white, and blue swirls of the Pepsi logo.

Investment in Venezuela, and especially foreign investment, had been decreasing over the previous few years, driven partially by Chávez's anti-capitalism rhetoric. Between when Chávez took power in 1999 and when I biked across the country nearly a decade later, 40 percent of industrial companies were shuttered. Nonetheless, because of high oil prices, Venezuela's gross domestic product had actually grown significantly in the previous few years. This growth, though, had led to more consumption, such as new cars or televisions, and not much to investment, such as new buildings.

———————

After thanking my silent driver, I lifted *del Fuego* from the bed of the truck, thudded its wheels against the pavement, and started biking across the city to the apartment of Tony, a man who'd offered me lodging via email.

As I biked, cars and buses, honking and barely moving, sat in bumper-to-bumper traffic. I squeezed between the vehicles, briefly took up an entire lane, and then retreated to the sidewalk. An ambulance tried to wade through the traffic, blasting its siren, but no cars cleared the way—instead they passed in front of the ambulance, cutting it off and jockeying for position. Biking on the sidewalk, *del Fuego* and I easily outpaced the emergency vehicle, its siren fading behind us. I felt shocked and disturbed knowing I was faster than an ambulance. *What does it say about a society that won't make way for an emergency vehicle?*

I soon arrived at the apartment building of my host for the evening. Tony was an athletic trainer and cycling enthusiast who'd once been married to a U.S. citizen; he'd lived in California for a decade before returning to Venezuela. Tony warmly invited me into his apartment, which had all the amenities of the first world and which he shared with his sister. I was glad to join them in a meal of salmon and fresh carrots.

Because I'd hitchhiked to Caracas, I arrived a few days earlier than my friend Tom was due. This gave me time to accomplish a number of errands, including a school visit, an interview with one of the national newspapers, and, perhaps most important, finding new malaria medicine, as the strain of malaria found in the Amazon basin is resistant to the anti-malaria drug I had been taking. Since reaching Central America, I had been taking a pill of Chloroquine once a week to protect myself from the disease. Malaria is present but uncommon through southern Mexico, Central America, and Colombia. It's more common—and resistant to the cheapest drugs—throughout the Amazon basin.

Jesús Castillo

Traffic in Caracas

So, the next morning I headed to a pharmacy that was just two stops away from Tony's apartment via Caracas's extensive subway system. I descended underground to the metro, where I found a white, well-kept train that was absolutely packed with passengers. Waiting for the next train, I realized I was the only foreigner in the crowd on the concrete platform. The second train was just as crowded, but I squeezed in nonetheless. When the train stopped at my station, I was pushed and pulled as a mass of people exited the car. I succeeded in forcefully extracting myself from the car, but quickly sensed something wasn't right. Reaching into my pocket I realized my wallet was gone.

Crap! I wanted to return to the car with fists flying, but, even if I could, I didn't know who I'd hit. With the train now gone, I tried to envision what I would have done differently, how I should have guarded myself on that subway. I'd lost forty dollars' worth of bolivars, and felt furious at the violation. That was three days' worth of travel money. Though the crime was minor—no documents, no physical harm—it was the first time I'd been robbed since losing my passport in California.

Walking on the street, I felt unfriendly eyes. The last time I'd felt this uncomfortable on a street was in Tegucigalpa, Honduras. Though now with no cash, I continued to the pharmacy anyway to see if they had the medicine. I walked quickly, hoping no other thieves would follow me. Had I had anything more to steal, it would have been almost as easy above ground as it had been below: men and women packed the city's narrow sidewalks. Above us towered ten-story concrete buildings whose windows had metal bars—even on the top floors. I felt vulnerable and scared, and wished I were safely on *del Fuego* riding in the street rather than walking on the sidewalk.

In planning the trip, I had most feared Central America and Colombia, but with Venezuela's crime on the rise, Caracas may have been the most dangerous city I visited during my entire journey. Its murder rate, at about six thousand per year, was then nearly nine times that of Bogotá. While I assumed most of the deadly violence occurs in the slums that rise up the mountains surrounding the city, I still felt wary downtown.

The pharmacy did not carry Mefloquine; I'd have to go to a hospital across town. Exasperated, I turned around and walked the mile and a half back to Tony's apartment. Welcome to Caracas.

I reemerged from the apartment with my trusty *del Fuego* and followed a busy road into a secondary valley that extended away from downtown. On the hillsides, small houses were stacked haphazardly, like blocks poorly placed by a three-year-old. Most of the structures were simple brown brick or gray concrete, though a few were painted white, green,

or turquoise. Climbing up every available bit of the hillside, the buildings appeared to have no roads or walkways between them.

After locking *del Fuego* to a fence in a guarded parking lot, and conversing with the security guard until I was confident the bike would be safe,
I entered the hospital in search of a nurse who worked on malaria and dengue fever. I found one sitting at a computer in a mostly empty room with
florescent lights and concrete floors. "We have that," she said, and handed
me a number of pills, offering them without requesting payment.

"Ahhh . . . I have those," I said, eyeing the Chloroquine. "I heard the
malaria here is resistant to that drug. Do you have Mefloquine?"

"No, you can't get that here," she said, "but this works."

I asked if she had information on where malaria was present.

"Yes," she said, and handed me a few binders on the subject, which
I spent the next twenty minutes with. Though there are about fifty
thousand cases of malaria each year in Venezuela, few are fatal, as the
strains in South America are far milder than those in Africa. Almost all
of Venezuela's cases of malaria are contracted in the southern, unpopulated part of the country, and apparently only a small percent of those
cases are resistant to the drug I was carrying. The disease had been
mostly eradicated from large parts of Venezuela in the middle of the
twentieth century due to the spraying of DDT, but for unknown reasons it had made a comeback in the last fifteen years.

The nurse also gave me information on dengue fever, another ailment transmitted by mosquitoes. Dengue is far more common; Caracas
reports about fifty cases every week.

Since mosquitoes survive better in warmer weather, they can spread
more disease in warmer climates. The major cities in Andean nations
were originally built in the cool mountains partially to avoid such tropical scourges—Caracas sits at 3,000 feet, Bogotá at nearly 9,000 feet, and
Medellín at 5,000 feet. Because of climate change, though, malaria already
appears to be creeping up the mountains. The World Health Organization
estimated that, in 2001, global warming was responsible for over 5 percent
of malaria and dengue cases in a number of countries.

Of course, malaria and dengue fever aren't the only health risks of a
warm climate. Hot weather increases mortality. During the summer of
2003 about thirty-five thousand people, mostly in France, died prematurely from an unusually strong heat wave. And a surprising number of

other diseases are more common in warmer weather. Diarrhea outbreaks are more likely in warmer years; the World Health Organization estimated that worldwide cases of diarrhea were 2 percent higher in the year 2000 than they would have been without global warming. Even cases of food poisoning are higher in warmer climates.

I sometimes hesitate to argue that climate change will seriously affect human health, largely because public health systems play a more significant role in people's well-being and can potentially offset most of the negative impacts of climate change. For example, the elderly in France could have survived the 2003 heat waves if they'd had air-conditioning. Malaria can be controlled through medicine and spraying for mosquitoes. And some deaths will be avoided through warmer winters, although that number will likely be fewer than the additional deaths caused by warmer summers. But all this assumes people have ready access to necessary preventatives or treatment. Those who live on the margins—the old, the young, the sick, the poor—will have a harder time adapting. According to the World Health Organization, existing climate change—the fact that the Earth's temperature is already 1 degree Fahrenheit warmer—is already causing one hundred fifty thousand people to die prematurely every year.

DAY 253
July 15—Tom Arrives

My friend Tom arrived at the airport with his mountain bike in a cardboard box. His hair was a buzz cut—the preferred cheap haircut for a graduate student—and he wore sandals, a simple white T-shirt, and a big smile. This was the Tom with whom I'd biked from Portland to Palo Alto a little over six years earlier. That trip had been Tom's idea; I would never have attempted it if he hadn't thought it possible.

"Welcome to Venezuela," I said laughing at Tom and giving him a hug.

"Thanks, man. Let's bike across it." He said this with the same joking confidence he'd used when he suggested that first bike trip back to college.

Now Tom was in the fourth year of his PhD program in physics. He'd decided he could join me in Venezuela because his advisor was on vacation for two weeks, and thus wouldn't notice his absence. Tom had also recently gotten engaged to his girlfriend. (He'd emailed me when

I was in Guatemala saying, "You have until Panama to object to me proposing to Lauren.")

So we had much to talk about on the taxi ride from the airport. I had been on the road for eight months. Other friends and acquaintances were transitioning from their twenties to their thirties, finding more serious jobs, becoming engaged—I'd missed the wedding and bachelor party of another good friend during my time away. For the first time I felt I was missing out by taking this seventeen-month vacation from my regular life. I wasn't making "progress" the way one is supposed to, the way my peers were. While normally I embraced the concept of forging my own path, Tom's stories made me realize more acutely all I was giving up during these travels.

Tom had brought with him two books about Venezuela—one critical of Chávez and one supportive—and he was eager to see the country for himself. "It's so weird," he said. "It's like a democracy that isn't."

We stayed with Esteban, a friend of Tom's father. An executive at a telecommunications firm, Esteban lived in a wealthy neighborhood high in the valley, had sent his daughters to college in the U.S., and spoke perfect English. When we asked him about Chávez, he stated the president was hurting the private sector, and slowing investment. "He's trying to turn this country into a second Cuba."

"What about the programs he's instituting to help the poor?" I asked.

"Yes, those," he said. "They're just to buy the vote."

———

Tom and I took a day to explore the town before leaving Caracas. As we walked through downtown on a Saturday, traffic was calmer than on the weekdays, and we found a street blocked off for a parade of protestors. They appeared peaceful, so we walked alongside, and Tom asked one of them why they were marching.

"We're protesting Chávez," answered the man, who was about our age. "He's squashing all dissent."

"How so? He seems to let you march."

"Yes, well, for one, remember the recall election in 2004?" The Venezuelan constitution allows for a recall election if a sufficient number of signatures are collected demanding such a vote. In 2004, two million people signed such a petition, and a nationwide election was held.

Chávez won handily. "Everyone who signed the recall petition has been blacklisted from government jobs," the man said. "Either you support Chávez, or you're kicked out. And the only new jobs these days are government jobs, so there is no place in this country for us."

"But wasn't Chávez democratically elected?"

"Yes, but he's screwing with the government. The Supreme Court ruled against him, so he has packed it with more judges favorable to him. It's all a sham. And he shouldn't use our votes against us—our votes are supposed to be confidential."

"What about the missions?"

"They aren't helping."

After a few blocks the march stopped; men and women raised Venezuelan flags while speakers belted out Venezuela's national anthem, a stately symphony of brass and woodwind instruments with a slow, marked rhythm.

The next day Tom and I headed out from Caracas, riding east along the coast accompanied by soaring temperatures. Tom pedaled a red aluminum Schwinn mountain bike with front shocks and skinny road tires; he also carried only two rear panniers—as one needs few warm clothes in the tropics—so his bike was far lighter than the steel *del Fuego* and its three saddlebags. Yet my many months of riding had made me strong, and I maintained his speed despite my extra load.

The miles ticked off easily as Tom and I swapped stories, though it was also a little strange, riding with a companion again. I was aware how Tom buffered me from the surroundings, almost providing a barrier between Venezuela and me. We were our own little bubble exploring the country, whereas riding alone forced me to interact more with local people, to experience the scenery passing me more mindfully. For that reason I was glad I did most of the trip alone, but that didn't stop riding with Tom from being a welcome break. Sharing this time with a real friend helped recharge me, and prepared me for the next few months when I'd again be pedaling alone.

Riding along the coast, we passed more than one roadside billboard displaying a large image of President Chávez. Chávez usually had his arm around the governor of the local region, the billboard announcing the construction of a bridge or road. In one small town a colorful mural depicted Simón Bolívar and Hugo Chávez and read, LONG LIVE TWENTY-FIRST CENTURY SOCIALISM. NO TO SAVAGE CAPITALISM. Another mural read EVERY DAY 30,000 CHILDREN IN THE WORLD DIE BECAUSE OF HUNGER CAUSED BY CAPITALISM. WHO SAYS THAT CAPITALISM IS PRETTY? Tom asked, "Do they have propaganda like this in other countries?"

"Not like this." It was true—no country came close to Venezuela in marketing its own government. The billboards always left me feeling uneasy. Why should a state have to advertise so intensely to its citizens?

DAY 258
July 20—Santa Fe

After three days of riding, we reached the small town of Santa Fe, which we'd heard had good beaches and beautiful coral reefs. We took the next day off to go snorkeling on the coast, and spent a night at a fire station. Then we turned inland, riding south, finally beginning the long southward push across South America. I felt the beginnings of another cold coming on—the second time in less than a month. But, once again, I couldn't afford to take time off to rest—we were on a tight schedule, as Tom had to catch a flight out of Puerto Ordaz.

Rather than take a day off, Tom proposed a solution. We loaded all our panniers onto *del Fuego*, whose front and rear rack could hold five bags, and we switched bikes. So, not only did I get to limply ride his lighter aluminum Schwinn, I drafted off him as well. It felt odd to be separated from *del Fuego* and yet to still see him, continuing ahead of me. I wonder how *del Fuego* felt.

After a long lunch break, my cold gradually improving as we continued south, we arrived at Los Llanos, the great plains Venezuela shares with Colombia, where tall grasses stretched to the horizon. This region annually experiences a heavy rainy season followed by an intense dry season. These dramatic seasons, combined with the infertility of the soil, have resulted in wide prairies instead of forest.

As the sun dropped into the grassy horizon, we searched for a campsite along the road. We soon encountered a lone one-armed man, who introduced himself as Gregorio and told us we could camp on his property.

Once our tents were up, Gregorio walked out to talk with us. He was shirtless, revealing his slight, skinny frame, and in his left arm he carried a machete. After thanking him for letting us camp, I asked, "Are you going to farm this land?"

"Yes, I just bought it." The land was sandy and covered with short brush. He explained he'd bought it with money from his old job. "I worked at a concrete factory. A machine cut off my arm, and the company gave me money, which I used to buy this land." He seemed somehow cheerful as he spoke. I asked him how common accidents are. "It happens all the time," he replied. We then listened to him talk on and on about how Chávez was helping the country rise up.

"Venezuela won't succeed overnight," he said. "We need a long time to build ourselves up as a country." As he spoke, he occasionally waved the machete for emphasis. "We will improve little by little." He repeated this phrase a few times as if repeating a talking point. "Chávez has the Bolivarian Dream."

"How has Chávez helped you?" I asked.

"Health care is better, and now education is free." He said this smiling.

But the man also had only one arm, his land sat fallow, and it was unclear if he'd be able to farm the sandy soil. As we took down camp the following day, Tom and I debated whether he'd be able to grow any crops. I don't think we came to a decisive verdict either way.

After another day's ride we made it to the city of Puerto Ordaz, where we shared a hotel room for Tom's last night. "Here, you can have the last few thousand of my bolivars," Tom said, handing me his currency. "I wonder how Simón Bolívar would feel about his little soldiers being worth so little," he joked, looking at the fistful of bills worth only a few U.S. dollars, the result of inflation over many years. "What do *you* think is the matter with this country?" he asked.

I remarked that Venezuela was somehow less pleasant to cross than other countries I'd cycled. "One reason is there are more cars here. The cities are just less enjoyable." Also, with the exception of highways in central

Mexico, only in Venezuela had I cycled busy roads in the countryside.

"I also just don't feel safe going out at night," Tom said.

"That's true in many Latin American cities, but especially here."

Tom made a few joking remarks about how he wished he had joined me "in a better country" or at least continued with me on the road south of there, which was the only route into Brazil. "That road looks amazing," he said, looking at the path leading to the center of the continent.

"Well, it's tough to bike a road like that and also work toward a PhD." Tom nodded, and we didn't say much more. The following morning, Tom took a taxi to the airport and returned home.

DAY 265
July 27—La Gran Sabana

Southern Venezuela was mostly uninhabited. Without population centers to visit or a riding partner to talk to, I biked all day, departing camp at sunrise and riding until dark, stopping only to refill on food and water. I listened to playlists on my iPod for perhaps the thirtieth time, letting the music push my legs forward. I was glad to be far from Venezuela's population centers, and far from oil production centers, although occasional roadside billboards still reminded me of Chávez's rule.

A spectacular campsite in La Gran Sabana

The road departed Los Llanos, cut through dense forest, and then climbed steeply, emerging from the jungle to another grassland, La Gran Sabana, a few thousand feet above sea level. The "great savanna" had topography unlike anywhere I'd ever visited, and I was fortunate it would take me a few days to get through it. The land was a series of terraces with steep drops between them, giving rise to odd flat-topped mountains and large waterfalls. (Angel Falls, at 3,000 feet the world's tallest waterfall, was nearby, but too far off route to visit.) I passed numerous impressive cascades, the water dropping from one terrace to the next. As the rainy season had just begun, afternoon rains occasionally pelted me, and the grass radiated green. In the afternoon I jumped in a river below a water-fall, and then at night set up my tent at the top of one of the terraces, from which I marveled at the sweeping view across the land.

At midday about fifty miles short of the Brazilian border, I stopped to pick up supplies in the small village of San Francisco de Yuruani. After purchasing pasta and a can of tuna, I chatted with the storeowner, who was reading over what looked like math homework. I asked what he was reading.

"Studying. I'm in Mission Ribas." Mission Ribas provides remedial education, helping adults finish a high school equivalency degree.

"How does Mission Ribas work?"

"Every weeknight except Friday we have a three-hour class."

"What happens if you don't go?"

"Well, we receive a stipend to attend." Paying students to go to class certainly encourages attendance.

He then showed me how he was learning order of operation in math—whether to multiply or add first when given a basic equation. He then asked me to help with his geography homework and opened a map—he was having trouble reading latitude and longitude.

"How have these classes helped you?"

"I can read now—I couldn't do that before."

He used to be illiterate, and now he can read. The thought sent a shiver down my spine, and I almost wanted to cry. *How different would the world be if I couldn't read?* What I had read of the missions—mostly in *The Economist* magazine or *The New York Times*—expressed doubt over their effectiveness, labeling them as not real programs, as just political payoffs for the poor. Also, nearly every person I'd spoken to in Venezuela's middle or upper class had discredited these programs—which probably says more about class divisions than it does the programs' effectiveness. Yet here was someone who had indeed benefitted, greatly: he could not read before, and now he could.

"How about schools for the young. Are they improved?" I asked him.

"Yes, the government now pays for lunch, so more students go to school."

Introducing himself as Fabian, the man invited me to join him and his wife, Adelicia, for lunch: chicken stew with potatoes and rice. During our meal they showed me the pile of readers that constituted the lesson plan for Mission Ribas. The reader for one of the classes was a book called *Human Rights,* which appeared to be propaganda for the government; another book was *Our Chávez,* a collection of essays about the virtues of the president. I helped Fabian and Adelicia a bit with their English homework and attempted to teach the rule about the silent *E* at the end of many words. I then asked about Chávez.

"He's the best president in South America," Adelicia said. Her husband nodded in agreement.

———

Back on the road, I was struck by the serendipity of this encounter. This lunch with Adelicia and Fabian was one of the few moments where I felt like Venezuela actually worked well for its people. But, then again, if Chávez's policies were helping, why was I leaving the country with a visceral feeling that the country was headed toward collapse? I mentally ticked off my evidence: the congested cities, the soaring crime rate, the faltering non-oil segment of the economy, oil engineers describing poor management—and the fact that everyone in the middle and upper classes detested Chávez. I found myself thinking of César, who spent his money from working on the oilrig on beer and a television instead of improving his family's house—essentially wasting his personal oil wealth. I thought too of the one-armed Gregorio, waving his machete and explaining how he was going to grow crops on the sandy ground, crops that could easily fail.

When I bike across a country, I take a non-scientific survey of the nation's mood. It's a subjective collection of experiences—the number of new buildings, the smiles on the streets, the ease of travel. In Colombia, I perceived a swell of pride and growth. People were excited that their country was improving. Colombians may still be dying in armed conflict, but today is better than yesterday was, and tomorrow will be even better. In Venezuela, in contrast, I felt like the country had both a physical disease and a psychological disorder.

While Chávez may have been to blame for Venezuela's precarious political and economic state, so is the so-called "curse of oil." When I'd asked various people my question about Venezuela and oil, I had expected the answers to be more nuanced—I never thought oil engineers would tell me oil is bad for the country. In thinking back on my time in Venezuela, and picturing the congestion, the bumper-to-bumper traffic, the inflation, the slums, the crime, the class division, the human discord, I simply felt ill. I'm left with the strange feeling that, if we found a replacement for oil, not only would we benefit, but Venezuela would be better off as well.

Excited to reach another country, I continued south, following the only paved highway into Brazil and the Amazon basin.

BOLIVARIAN REPUBLIC OF VENEZUELA
July 5–August 3
Presentations: 1
Flat tires: 6
Miles: 1,188
Trip odometer: 8,420 miles

PART IV
BRAZIL, PERU, AND BOLIVIA 🚲
The Amazon to the Andes

The original inspiration for my trip was not to save the planet, but to seek adventure—and the bicycle was merely my preferred mode of transport. Yet, in retrospect, I consider an Earth-crossing biking adventure intrinsically environmental. There's something spiritual about traversing a large swath of the planet under one's own power. By arduously pedaling over hills and across plains, through the rain and wind and heat, I felt more connected to the Earth, or at least I felt there was less of a distinction between me and the planet we live on. And by pedaling across more than 90 degrees of latitude, "the Earth" was no longer an abstract thing, but rather a tangible place with clear dimensions.

My adventure was most in step with the natural world in the Amazon Basin, where I was the farthest from civilization. Twice I encountered areas where roads didn't exist, forcing me to find alternate transportation—first up the Amazon River, which I traveled by boat, and then leaving the basin and crossing into the Andes, where I continued on foot. For two months I visited towns accessible only by boat, passed through indigenous reserves where the people still hunt with bow and arrow, and met individuals who told me they hadn't seen another "gringo" in two decades. I felt I had discovered the far corners of the Earth that my peers rarely see, and I loved it. But I also especially enjoyed arriving in such places under my own power, an act that for me physically connected these remote regions with other, more mundane places—random cities, or even my home in California.

My crossing of the basin and climb into the mountains wasn't as adventurous as it had been for early European travelers—and it was perhaps a thousand times less adventurous then the travels of the first Native Americans to reach the basin, who probably arrived some ten thousand years ago and had to figure out how to live within the jungle. Yet I still internalized the forest's vastness, feeling its heat and experiencing the thick buzzing of life within. And while this area may still be one of the most remote parts of the globe, even this region is changing in response to human civilization.

Brazil route

15 BRAZIL
A Forest in Flux

Brazil population: 199 million
Annual per capita GDP: $11,900
Annual per capita CO$_2$ emissions from fossil fuels: 2.2 tons

DAY 273
August 4—Boa Vista

At the border, along a seemingly abandoned stretch of road, I was met by Brazil's green, yellow, and blue flag. In white letters along its middle read ORDEM E PROGRESSO, Brazil's official motto: "Order and Progress" in Portuguese. In front of me, though, there was little sign of order or progress— or civilization for that matter. Beyond the customs building, an empty mix of shrubs and grasslands stretched southward. Although Brazil is South America's most populous country (one out of every two South Americans lives there), you couldn't tell from the border. Most Brazilians live along the Atlantic coast, far away from where *del Fuego* and I found ourselves.

The road I would follow, BR-174, is the only paved highway through a vast swath of the Amazon. It's as if the entire United States east of the Mississippi were covered in tropical rainforest and had only one paved road running north–south. The jungle does have a few dirt roads, but some are nearly impassable in the rainy season, and none would take me all the way west to the Andes. So to reach Peru I'd have to bike seven hundred miles south to the Amazon River and then take a boat a thousand miles upstream.

On a road descending from the high grasslands of La Gran Sabana, I rolled through jungle for a few hours. Gradually the terrain changed to wet grasslands, where occasional hills rose like islands above an otherwise flat plain of grass and shrubs—plus a few palm trees growing in standing pools of water. The grasslands made up just a small portion of the basin; the vast forest awaited me ahead. Tall clouds threatened rain but then receded.

I spent the first night camped next to a lonely home, the only structure for about ten miles in either direction. I soon discovered that, though Spanish may be similar enough to Brazilian Portuguese to ask whether or not I could pitch a tent, it was not close enough for me to understand basic questions like: "How long have you been biking?" and "Where are you from?" (I replied, in Spanish, "ten months" and "California," assuming those were the questions being asked.) I was thankful that night for the mesh in my tent keeping the mosquitoes out. Although I had finally acquired the pills necessary to defend against the malaria of a mosquito bite, I had no remedy for its itch.

———————

After another day of riding, with my arms and face thick with sweat and sunblock, I headed into Boa Vista, a city of about one hundred thousand people 140 miles from the Venezuelan border, the largest settlement for a week of travel in either direction. Entering the town I was confronted by Portuguese signs displaying words that were almost Spanish: I took ADVOGADOS to mean *abogados* (lawyers), JÓIAS to mean *joyas* (jewelry), and CORREIOS to mean *correos* (mail).

Fortunately, the phrase "Where is the fire station?" is almost identical in Portuguese and Spanish, though in Portuguese they are *bombeiros,* not *bomberos.* Following the gestures of a series of pedestrians, and navigating the city's short buildings and light traffic, before long I arrived at a red and white building. When I presented my stack of firefighter patches and my *Bomberos de Costa Rica* uniform, I was invited into the dormitory. I was happy; they had a shower, an air-conditioned room, and an extra mattress. Though I understood little of what the *bombeiros* said, they somewhat understood me when I spoke slowly.

In the morning, Paolo, a skinny *bombeiro* with short curly dark hair, helped me conjugate basic Portuguese verbs. He showed me how to pronounce the new letter *Ç,* the *RR,* and the nasal *Ã,* which required stretching my neck and attempting to breathe through my nose. As we practiced, phrases of Portuguese gradually untangled themselves in my mind, revealing themselves as altered Spanish.

Paolo then said something that sounded like, "bla bla bla television bla bla."

"Television?"

"Yes, tomorrow."

———————

The following day, a reporter and cameraman met *del Fuego* and me on the banks of the Río Branco, a half-mile-wide river. On the far bank was a thin strip of trees, and in the sky clouds were growing tall, absorbing moisture from the jungle. First the cameraman filmed me riding along the river, after which I was to speak to the reporter.

I can't say that the interview went well. The reporter held the microphone limp and rolled her eyes with frustration when she realized I couldn't understand what she said. If they decided to air the interview at all, they'd have to edit out almost everything—except the footage of me biking along the river bank.

DAY 277
August 8—BR-174

I continued south, now very near the equator. Sweat beaded on my skin, forcing me to frequently reapply sunblock. Though the grasslands gave way to rainforest, the forest rarely reached the highway—for a hundred yards in each direction, the roadside had been cleared for cattle grazing or farming. The lack of foliage along the road, however, offered a peek into the habitat within, especially the birds: toucans, parrots, smaller yellow birds, small chirping black jays that jumped on the power lines, woodpeckers pecking the posts holding the lines, and black vultures perched on fence posts. An iguana-like lizard scurried into the grass next to me, and a long black snake crossed the road ahead of me.

Most of the Amazon's deforestation occurs along the southeastern edge of the forest, a thousand miles from where I was. In the northern portion of the basin, where I was riding, the highway marks the only line of felled trees for hundreds of miles to the east and west.

From the seat of *del Fuego*, I gazed at a herd of cattle standing between the jungle and me. Cattle and other livestock are responsible for much of the Amazon's deforestation, and it's estimated that nearly half of the world's deforestation is directly or indirectly the result of livestock—because forests are cut to make way for grazing land and for crops to feed the animals. Also, given that cattle and other animals produce large quantities of methane (a surprising 5 percent of global greenhouse gas pollution comes from the belching and flatulence of cattle and other ruminants), and that the farming practices supporting livestock produce the greenhouse gas nitrous oxide, livestock are responsible for nearly one-fifth of global greenhouse gas pollution—more pollution than all cars, planes, trains, and ships combined.

This fact is somewhat shocking, and stating it often produces incredulous stares—it's hard to believe that one could reduce more pollution by giving up beef than by switching to a hybrid car. Perhaps we don't talk about this because it's easier to imagine completely reinventing our energy system than to envision the world voluntarily eating less meat.

Cattle along highway 174

We have another challenge as well. Just as people in poorer nations consume less energy, they also consume less meat, and as incomes across the world increase, demand for meat is expected to grow. The average American or Western European eats a thousand calories a day from animal sources, while the average person in the world consumes fewer than five hundred. If people become wealthier, meat consumption will rise, worsening climate change. We will either have to produce meat more efficiently, or consume much less of it.

The cattle and farmland, with the forest just out of reach, would be my roadside companions for nearly 200 miles. Farther south, the highway passes through the Waimiri-Atroari Indian Reserve—land set aside for Indians who apparently still survive by hunting and gathering. I had been looking forward to pedaling through the seventy-mile stretch of untouched jungle. But I would have to do so in a single day; camping wasn't allowed, and I'd heard that two German cyclists had recently upset the natives by pitching a tent in the middle of the reserve.

"They will kill you and eat you!" exclaimed a roundish middle-aged man at the Petrobras gas station. "You can't bike through the Waimiri Indian Reserve!" At least, that's what I thought he said—my Portuguese had improved, but I wasn't yet conversant. The man then added something I couldn't understand and made clawing motions with his hands as he spoke. Was he saying that the jaguars would eat me if the natives didn't?

"I don't believe you," I said in half Spanish, half Portuguese. "They don't eat people here." I assumed he was prejudiced against indigenous people and merely repeated falsehoods he'd heard elsewhere.

"Nooo," he replied emphatically, and then pointed to a woman standing behind him at the checkout counter of the tiny gas station. "They ate her husband." The woman solemnly nodded her head in agreement.

I looked down the road a half mile to the reserve's entrance, suddenly feeling something I hadn't felt in a long time: fear. The roadsides were cleared on either side of the highway. Then, at the reserve, the cleared land was met by a wall of dark trees, and the road disappeared into a small break in the foliage as if the highway were entering a gate into a fortress. I found myself envisioning the scene in *Raiders of the Lost Ark* where a tribe of headhunters chased Indiana Jones through the jungle. That vision couldn't be accurate, could it?

I pedaled a half mile to the edge of the forest, where a police check-point stopped cars. I approached an officer and asked about the gas attendant's warning. "Those incidents are way in the past," the officer replied. "Those conflicts were with farmers, not tourists." Indeed, after seeing how farmers had devoured the forest outside the reserve, I could understand why the Waimiri would be upset. I'm sure that disputes over the boundary between forest and grazing land has caused more than one conflict.

The officer added something I didn't understand. I asked him to repeat himself and speak more slowly, and I thought he said, "Stick to the road and you'll be fine. Also, the Waimiri ask that you don't take pictures or stop for any reason." As it was already late in the afternoon, I biked back to the small town—a line of buildings along the main road—and got a hotel room for the evening.

DAY 281
August 12—Waimiri-Atroari Indian Reserve

At dawn I biked toward the forest. The road was wet from a night-time storm, and my tires hummed along as they sprayed water into *del Fuego*'s fenders. When I reached the entrance to the forest I found myself beneath sagging foliage thick with moisture from the recent rains. Each branch seemed to have a slightly different set of leaves—some rounded, some ruffled, some long and skinny like palm fronds. Sounds filled the forest: insects buzzed, clicked, and rattled. Five small brown monkeys slowly crossed the road and then scurried into the trees. I saw no signs of human settlement.

Heeding the police officer's advice, I continued on, pedaling at a pace I could maintain all day. I found myself staring at the tree branches, wishing each tree was labeled with its name. The biodiversity of the Amazon jungle is among the world's highest. In just over two acres of this rainforest, a biologist once recorded 425 different species of trees. Another biologist found 43 different species of ants living in a single tree. Because the forest is so large—accounting for half the world's tropical rainforest—it's estimated that one out of every ten species on Earth lives within this basin.

Some climate models predict that global warming will dry out the Amazon Basin, and rainfall will decrease by half. Under such a scenario, standing trees would die, dry out, and burn, blackening the sky. The westernmost portions of the forest might survive, but the majority of the Amazon would give way to savannah or desert. If those climate models are correct, sometime after 2050 the Amazon will succumb to a wave of drought and fire. And although a few more recent studies have suggested this isn't highly likely, it is still possible.

Of course, if that were to happen, the burning of the Amazon would release even more carbon dioxide into the atmosphere, which would in turn worsen global warming. This result is perhaps the greatest concern—that a little bit of warming will result in so-called "positive feedback" loops that will accelerate warming. The climate system has many possible positive feedback loops. For instance, an enormous amount of organic material is locked away in the tundra of Siberia and northern North America. Warmer temperatures would allow soil microbes to digest this organic material, which would emit huge amounts of methane and carbon dioxide to the atmosphere. A warmer climate will also likely increase forest fires, which would put more carbon dioxide into the air.

In my presentations, I often speak about the climate as if we knew exactly how much the planet will warm based on how much greenhouse gases we put into the atmosphere. Yet there is huge uncertainty, and it's very possible that warming the planet just a little bit will result in positive feedbacks that would ultimately warm the planet a great deal. Partly for this reason many scientists argue that we need to keep the Earth's average warming under 2 degrees Celsius (3.5 degrees Fahrenheit); above such warming, it might be more difficult to slow climate change.

Riding through the Amazon, I tried to imagine the forest drying, turning brown, and being replaced by desert shrubs and grasses. Should that happen, both claimants to this region—the Waimiri Indians and the farmers deforesting the roadside—would be forced to abandon their ways of life.

Afternoon clouds formed above me as I biked, and rain pelted the forest with large cool drops. Though I stopped to put rain covers on my saddlebags, I remounted quickly. *Keep going! Don't let anyone see me!* I didn't want to offend the Waimiri. The storm soon dissipated and the sun reemerged, now almost directly overhead. The equatorial sun beat on the wet pavement, evaporating the water and turning the ride into a steam bath. I took my camera out of my handlebar bag and took a quick picture while pedaling—hoping no one saw me.

Brazil highway 174

After a full day of riding—seeing no signs of people or settlement for seventy miles—I reached the end of the indigenous reserve, marked by a sign and the reemergence of cleared roadside. Exhausted, I dismounted *del Fuego*. To the west of the road, a gravel driveway led to a small circular wooden hut with a palm-leaf roof, located at the edge of the reserve. A small sign indicated the building was property of the Waimiri Indians. Disregarding the advice of the man at the gas station, and feeling emboldened by my successful crossing of the reserve, I walked my bike down the driveway to see who lived there.

I was a bit nervous—I had just biked without stopping for seven hours because I'd been told to. But now I wanted to meet this tribe, wondering what they were like. It's tempting to believe that the native people in the Amazon, the headhunters of childhood stories, have a lifestyle that hasn't changed in millennia. But, unfortunately, we lack reliable knowledge about the lives of people before the Spaniards and Portuguese arrived. Archeology suggests that a substantial population of natives lived in the Amazon jungle until the waves of smallpox and other European diseases spread across the continent, and today the forest natives are mere remnants of a past civilization, like tribes in Europe after the fall of Rome. Some archeologists suggest that Indian populations greatly influenced the forests, selectively promoting fruit trees, and that a number of the trees I saw in the forest were actually "domesticated." Humans have likely been shaping this forest for millennia—the question now is how we will shape the forest in the next few decades.

I walked through the open door of the hut. Standing inside was a man with an angular face, maybe twenty years old, wearing a plain T-shirt and slacks. On the wall hung a hammock and a series of bows, arrows, and spears, all carved out of wood. In the middle of the room, on a series of wooden racks, dangled necklaces and other ornaments made of crocodile teeth or large black beads. The items were all neatly arranged and displayed, and each had a plain paper tag penciled with a number. I laughed: they didn't want to eat me; they wanted to sell me indigenous souvenirs.

I communicated with the man in a mix of Spanish and Portuguese. He spoke limited Portuguese, and we conversed slowly, waving with our hands and pointing. "How much for the hammock?" I asked, eyeing the thick brownish twine.

"Fifty reals," he said, or about twenty-five dollars.

"How is it made?"

"From fibers in a tree."

"How long to make a hammock like this?"

"A woman worked for a month."

Fascinated, I continued to ask questions, communicating through short Portuguese sentences and hand gestures. He seemed to say that he hunted with bow and arrow, didn't drink Coca-Cola, and didn't watch television. Perhaps these tribes were indeed isolated from the rest of the world. Of course, the man also spoke Portuguese and wore a T-shirt, so he must have spent some time outside the forest.

"Is it okay to bike across the reserve?" I asked.

"Yes, but we don't like it," he said flatly, and then glanced uncomfortably at my bike. I decided not to ask whether I could camp there— or about cannibalism.

I purchased the twine hammock and stowed it in my extra stuff sack, strapping it to the top of my rear rack, and then continued south down the highway. After half a mile, I encountered a small wooden building that, in addition to serving as a farmer's home, housed a small store selling soda, Manaus coffee, and ice cream flavored with açaí, a sweet purple fruit. The owner of the store let me camp at the edge of the property, and gave me a small Brazilian flag to affix to the back of *del Fuego*. The number of Brazilian flags I'd seen had impressed me, suggesting the patriotism in Brazil was greater than that of any other country I'd visited. In many ways it reminded me of parts of my own country—Brazil, like the United States, is a large, patriotic country.

The man also let me fill my water bladder with cool water from his refrigerator, which he said was filtered and safe to drink. As I went to sleep, resting on top of my sleeping bag in my boxers, I laid the water bladder across my stomach to keep me cool in the hot night.

DAY 283
August 14—Manaus

After two more days of riding, *del Fuego* and I approached Manaus, a metropolis of 1.5 million that was made rich in the early parts of the twentieth century from selling rubber trees (before synthetic rubber caused the economy to collapse). It had been a month since I'd encountered real traffic, and in approaching this bustling city I got a rush of adrenaline I hadn't felt since Caracas. I was also reminded of the harsh realities of these cities, and their horrible inequities. Next to a slum where cows grazed between haphazardly placed buildings, and where plastic bags and other trash lay uncollected, I passed a new development under construction. A sign advertised a future swimming pool for the modern looking development that was to be named Forest Hill. The sign also boasted there would be a 2-meter (6.5-foot) wall. It felt somewhat invigorating to see there was enough economic activity to build such modern structures; it felt less invigorating to learn it needed to be protected by walls. Indeed, Brazil has not only one of the largest gaps between rich and poor in the world, its gap is considered high even by Latin American standards. No matter how much I cycle through such inequity, I doubt I'll ever get used to it.

Dodging cars and riding through the city I eventually arrived at my destination, the Instituto Nacional de Pesquisas da Amazônia (INPA) or National Institute of Amazon Research, where I had connections through my former work. There I met Fabricio, a tall graduate student roughly my age, who had agreed to host me while in Manaus. We walked a few blocks from the research center to the apartment he shared with another graduate student, Alex. Speaking perfect English, they told me I could stay on their couch as long as I wanted. After over six hundred miles of biking across Brazil and the Amazon, I was relieved to have a place to call home for a few days before taking a boat upriver.

The following day I visited INPA. It felt like any major research center: a mix of walkways, parking lots, and manicured bushes separating two- and three-story office buildings and laboratories, all of which reminded me of my old research job. Yet it was also clear we were in the Amazon—tall tropical trees arched over the buildings, the trees' thick roots fanning out from smooth trunks, insects constantly buzzing. Inside the building where Fabricio and Alex worked, fluorescent lights illuminated plain white walls and disorganized piles of books. And though I heard Portuguese spoken throughout, everyone replied to me in English.

I met and chatted with Theotonio, a short man with thin glasses and a neatly shaved head, who models how clouds form in the Amazon basin. Theotonio looked over a paper on deforestation in the Amazon with me.

"Here is the Brazilian forest in 1992, and in 2002. The dark gray is deforestation," he said, speaking his English sentences slowly and deliberately, pointing to images.

The much-publicized deforestation of the Amazon was hundreds of miles away, along the southeastern border of the forest. "A little over half of this deforestation is from cattle plantations, and about a third from small subsistence farmers. Now, look at what will happen," the scientist said, "if we continue deforesting at the current rate. This is 2033."

The dark gray advanced far into the forest, eliminating most of the southeastern edges, and the road I had just biked was buffered by a far wider swath of dark. "A quarter of the forest will be gone due to deforestation. And if this continues, by 2050, nearly half of the forest will be gone."

Theotonio then said, "To make matters worse, the deforestation might turn the southeastern Amazon into a desert. About half of the rain in the forest is recycled water. That is, the water falls, evaporates, and then falls again. The rangeland and cropland that replace the forest don't cause as much evaporation as the forest. Also, evaporation creates clouds, which help shade the region and keep it cooler. So less forest means fewer clouds, hotter weather, and less rain. The result is the deforested land would then be too dry for agriculture—so the farmers would move farther inland, cutting even more of the forest."

I usually don't talk much about deforestation when I give my presentations, choosing instead to focus on fossil fuel use. I do this partly because pollution from fossil fuel use is growing more quickly than pollution from deforestation, and partly because deforestation decreases with increasing wealth. (Also, the carbon released during deforestation can be recaptured if the forest is allowed to grow back, even though it might take decades to reverse the process.) Much of the world's deforestation occurs in poor tropical nations, partially driven by subsistence farmers who have no other economic options. If Brazil's economy grows, its deforestation will likely slow—while its fossil fuel use will grow.

I asked Theotonio, "Deforestation currently makes Brazil's total carbon dioxide emissions among the world's highest, right?"

"Yes, that's true, and over three-quarters of our emissions are from deforestation—but you know, even with this deforestation, we still put only one-sixth as much carbon dioxide into the atmosphere as the United States." He then added that a lot of that deforestation is for raising cattle, much of which is for export.

"Yes, I know, I know. To be honest, I'm often embarrassed to talk about this issue." I told Theotonio that the success of my outreach—appearing in national news in almost every country—had convinced me I should continue the project in my own country, which pollutes more than all of Latin America combined.

"That is good," he said, half of his mouth curling in a smile. "The U.S. is polluting the most. It's like, when you are in an elevator with a bunch of people, and one person just keeps on farting. That person needs to change what he eats."

I laughed, at first not sure how to respond. "You guys are farting too," I said.

"Yeah, but not nearly as much!"

―――――――

In the half decade since my bike trip, something remarkable has happened in Brazil: the deforestation rate has dropped dramatically. According to government statistics, from 2002 to 2004 about thirty thousand square miles of forest were cut down, about the area of South Carolina. A few years later, between 2010 and 2012, Brazilians felled only seven thousand square miles—a reduction of about three-quarters. That means that the projected deforestation Theotonio showed me was more dire than has played out—fortunately we are not currently on pace to cut down half the forest by 2050.

Experts debate the many possible reasons why exactly deforestation has slowed. For one, Brazil's currency has increased in value, making it less valuable for Brazilians to export agricultural products. Also, new regulation and enforcement of existing regulation—driven in part by increasing environmental awareness of the Brazilian population—has also had an effect. But all the same, deforestation has not stopped—every year there is less forest than the year before. We need to continue to slow this deforestation, as well as deforestation in other tropical regions of the world, namely Indonesia and parts of Africa, which continue unabated. But, at least in Brazil, the trend moves in the right direction.

―――――――

I spent a week in Manaus, using INPA's high-speed Internet to start planning my U.S. journey. Though I was by this point determined with my plan, I still harbored some reservations. Given how far I'd already pedaled, I wasn't sure I'd want to keep riding once I reached Argentina, but I also wanted to compare these faraway lands with my own country. Most important, I wanted to discuss climate change with my fellow major carbon users; it felt dishonest to ride across Latin America with my message but then not also do so in the place I called home.

Also, I wouldn't be riding the U.S. portion on my own—I had decided to bike the States with a climate advocate named Bill Bradlee. Bill had contacted me by email a month earlier, and we'd exchanged emails and had a few Skype conversations about a potential "Ride For Climate USA" project. Though I felt a bit funny agreeing to bike across the U.S. and spend many months with someone I'd never met, Bill's enthusiasm for organizing the trip convinced me to go ahead. Plus, Bill took on most of the organizational work while I proceeded through

South America, which was a great help. In the meantime we outlined the website, discussed our route, and devised an outreach strategy. I would be mostly out of money by the time I reached Argentina, so we also explored fundraising ideas.

While in Manaus, I gave a talk at INPA and a local university, and also appeared briefly on the news. Fabricio and Alex provided me with Heineken at night and Cheerios in the morning, and I slept comfortably on their couch despite the hot nights.

DAY 291
August 22—The Amazon River

Manaus lies where the Río Solimões and the Río Negro combine to form the Amazon, the world's mightiest river. I planned to follow the Río Solimões—the main fork of the Amazon River—upstream into Peru (where it confusingly changes its name back to "Amazon River"). After reviewing boat schedules and purchasing a ticket, I wheeled *del Fuego* to Manaus's port.

The city's shoreline appeared more like an ocean port than a riverfront, as the far shore was over a mile and a half away, and just upriver the Río Negro widened to over six miles across. To my right a crane unloaded tractor-trailer containers from an ocean-going ship. In front of me, moored to piers, were what looked like a series of similar rowboats attached to the dock of a pond, except each boat was three stories tall and about one hundred feet long. The bottom levels of the boats were filled with boxes and crates—goods to be transported up and down the river. The second level of each boat was open, with a few dozen hammocks hanging from the ceiling. Each top floor was an open deck except for the ship's bridge located in the front. Though all the boats were painted white, each had a different color trimming the decks' railings—most with green, blue, or red. These diesel-powered boats are the buses and the trucks of the Amazon River, transporting goods and people to the towns along many thousands of miles of different riverbanks.

After purchasing a large blue hammock from a store along the crowded streets near the waterfront (the hammock from the Waimiri wasn't comfortable, so I mailed it to my father as a gift), I successfully found a boat traveling upriver. For one hundred dollars, the *Manoel*

Monteiro II would take me six days to the border with Peru—plus three meals a day and space to hang my hammock. I boarded and locked *del Fuego* to a pipe on the bottom deck, next to egg cartons, soda bottles, and boxes of clothing. I climbed to the second level, where fifty hammocks hung from the ceiling, weaving colors of red, blue, orange, and yellow. Most passengers were Brazilians, and young children darted about the deck, yelling and laughing.

Manaus's port on the Río Negro

At dusk, the *Manoel Monteiro II*'s diesel engine powered the boat away from the dock. The *Monteiro* motored downstream for eight miles before meeting the Río Solimões and beginning the long journey upriver toward Peru. A breeze worked its way along the boat, cooling the deck, keeping the hammock-berths comfortable. I befriended a young Spanish-Colombian couple who had rented a small cabin onboard, and they let me stow my valuables with them while I slept on the deck with the other passengers.

I quickly fell into a daily rhythm on the *Manoel Monteiro II*. Wake up. Jog in circles on the upper deck—my body, so used to cycling long hours, couldn't sit all day. Take a shower and enjoy a breakfast of bread and coffee served at the stern of the boat. Lie in the hammock and read. I worked my way through a history of Latin America, a book on modern Latin American politics, and the fourth

Harry Potter book in Spanish. Then a lunch of beans and rice and bread and salad, followed by a nap, followed by more reading, followed by dinner of fish or beef stew, followed by sitting on the top deck as the day faded, conversing in Spanish with my new friends, looking across the river as the trees on the banks slowly passed by.

Hammocks on the *Manoel Monteiro II*

Looking back from the top deck of the *Manoel Monteiro II*

As we progressed, the banks of the river slowly drew closer together, as the river narrowed from immense (over four miles wide in some stretches) to impressive (a mere mile across). I felt awe over the vastness of the forest and river, and I saw why some people mistakenly believe that humans could never really harm nature, since the natural world is so big and we are so small. But to me what is truly amazing is that, even though the natural world is so huge, we *still* have the power to permanently alter it.

I also felt a relaxed sense of wonder that I was in the center of the continent floating up the Amazon River. I was so far from California, yet I could think back on my trip day by day, remembering every campsite and fire station, seeing in my mind the unbroken stretch of road, river, and sea connecting this point where I floated to my old house in Palo Alto. And though I was no longer traveling under my own power (and was even carried by a diesel engine), I was glad that the Amazon was part of the path leading to Tierra del Fuego. I was still making slow progress, similar to the speed of my bicycle, and it was the pace of that progress that helped connect the start and end of my journey.

The boat would soon drop me off in Peru, where I'd find a second boat to travel yet another week until I reached roads I could traverse on my own. That road on my map was a small squiggly line, likely an unimproved dirt road leading from the jungle to the heights of the Peruvian Andes.

BRAZIL
August 4–28
Presentations: 3
Flat tires: 3
Miles: 651
Trip odometer: 9,071 miles

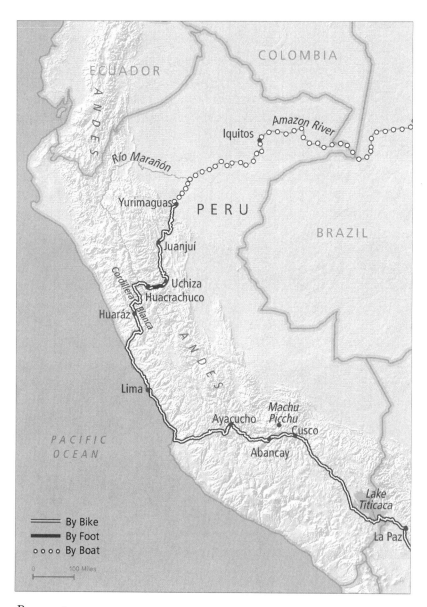

Peru route

16 PERU I
From Jungle to Mountaintops

Peru population: 30 million
Annual per capita GDP: $10,900
Annual per capita CO_2 emissions from fossil fuels: 2 tons

DAY 306
September 9—Juanjuí

After two weeks by boat and four days biking through the heat of the Peruvian Amazon on an exhausting, bumpy dirt road, I found myself defeated: the road I planned to take next didn't exist.

For the previous few months I had stared at the map and visualized following a particular small road into the Andes. You can usually tell which are the best roads for cycling from just looking at a map—the roads connect two inconsequential places, and are never the fastest route to major population centers. The road through Río Abiseo National Park had looked perfect. It would lead into the Andes and to roads that would bring me to the Cordillera Blanca, the world's tallest tropical mountain range, whose glaciated peaks I had dreamed of ever since I first planned my trip.

Unfortunately, at the tourism office in Juanjuí, a small town of about twenty thousand people at the edge of Río Abiseo National Park, I learned there was no such road. It wasn't even a trail. As to how one might traverse this area: "Some people bushwhacked through the forest," the tourism agent told me. "It took them a month." There was no way *del Fuego* and I could do that.

Of course, if I rode a week north or a week and a half south, I could follow a paved road across the mountains, but then I'd be bypassing the Cordillera Blanca, one of the highlights of South America. But there was possibly one more option—my map showed another crossing into the mountains just a few days south of where I was. It led from the town of Uchiza, in the jungle, to the village of Huacrachuco, in the Andes. Reaching this road would require biking south through Peru's coca-growing region on a stretch of road that the Lonely Planet guidebook warned one should not travel—on account of bandits. But then again, I didn't know whether to trust my map. What if I took the risk of riding through a dangerous region only to have the road from Uchiza to Huacrachuco also not exist? Disconcertingly, the tourism agent had no idea whether the road existed either—apparently he had never traveled more than a few days south of Juanjuí.

Frustrated, I took my now normal course of action: I went to the fire station in search of friends and advice. I found both at the Juanjuí fire station, a one-story building with an open-air garage, a gravel driveway, a small office, and a single-room dormitory. They had one fire truck that was only slightly larger than a pickup. I showed off my firefighter patches to the volunteer fire chief, Jerry, and after a short conversation he invited me to both stay the night and join him for lunch at a nearby restaurant.

A slightly pudgy man in his late forties, Jerry wore a button-down shirt, slacks, and sneakers. He worked as a doctor at the local hospital in addition to leading the volunteer force of half a dozen firefighters. Over a meal of *lomo saltado*—salty beef strips with rice and French fries—I showed Jerry my map and asked if the road from Uchiza to Huacrachuco existed.

Jerry shrugged. "I don't know," he said. My heart sank. Of course, Uchiza was a three-and-a-half-day bike ride away; as I had discovered on my trip, most people didn't know about roads more than half a day away from where they lived.

"Is it safe to bike to Uchiza from here?" I asked, wondering if it would be safe to even approach the hypothetical road.

"I don't think it's dangerous," he said, "but it is the coca-growing region." Jerry explained that, as the leaves from the coca plant are refined to make cocaine, the U.S. government flew helicopters over the area to find and eradicate coca plants.

"Does the eradication work?"

"No," Jerry said, taking another bite of his lunch. "The stuff grows everywhere." I felt my stomach clench. I doubted the coca-growing region could truly be safe, especially for a gringo from the States. Would the locals direct their anger at me, as my government funded the choppers?

As we talked, the dark clouds above the town collected into a thunderstorm—heavy rain pelted the restaurant's tin roof, and a light spray spattered through the open windows. A bright flash was immediately followed by a boom that shook us in our seats. Soon we heard the siren of the town's sole fire truck. The chief stared at me with surprise through his circular glasses—he should probably have been on that truck.

We hurried back to the three-room fire station. In the building's office, the chief tried to call another *bombero*'s cell phone, but he couldn't get through. Jerry searched the small office for the note that the secretary, also a *bombero*, should have left behind indicating where they went. As he looked for the note, the fire station's phone rang.

"Yes . . . okay . . . okay . . . we'll be right there." Jerry hung up and looked at me with wide eyes. "There's a second fire. Let's go." My heart pumped with excitement and a bit of fear. Though I'd stayed at almost two dozen fire stations, I never imagined I'd go to an actual fire.

I got on a motorcycle behind Jerry. He wasn't in uniform, and we had no equipment—the only fire truck was at the *other* emergency. The chief revved the engine and flipped on a loud siren. He looked back at me and smiled with pride: "I just installed this siren." He didn't explain, though, how it would put out a fire.

I grasped the handgrips by my thighs and we roared out, motoring over concrete and dirt roads, the rain now just a drizzle. At thirty miles an hour, we rode the fastest vehicle in town. Most of the light traffic we encountered was other motorcycles or three-wheeled motor-taxis, a hybrid vehicle that looked like a motorcycle in front and a small chariot in back. We passed the main plaza, lined with palm trees, and then drove by a park from which I could see green hills forming the edges of the Huallaga River Valley.

We arrived at a one-story home made of cinder blocks with a tile roof, finding a fire at the top of a palm tree in the family's backyard. Lightning had struck the skinny tree, which now burned like a candle with a small flame. Without equipment, the chief and I couldn't do anything other than watch the fire, and we stood in silence, staring at the display. I wondered what the residents thought of their fire department—responding to their call was a motorcycle with a siren, a doctor in slacks and a button-down shirt, and a tall American tourist with a digital camera.

The neighbors watched the fire through a hole in the brick wall between the two properties. "Hey gringo!" a teenager called to me curiously. "What are you doing here?"

"I'm biking across Peru," I said, and walked over to the hole in the brick wall, camera in hand.

"Really? Why?"

"Well, I'm raising awareness of global warming."

"Oh, you mean like the ozone layer."

"No . . . not exactly."

"So, what then?"

Bomberos of Juanjuí (Jerry is next to me)

I briefly explained the difference between global warming and the ozone layer while Jerry watched the drizzle put out the fire. I was surprised that even in remote parts of the Peruvian jungle people had heard of global warming—though I was no longer surprised to hear about the ozone.

The fire chief watched the fire attentively, focusing as if he were guiding each raindrop onto the flame. When the fire was finally out, he looked at me. "Okay, our work is done here. Let's go."

We returned to the fire station to find the secretary and two other *bomberos* recently returned from a fire on the far side of town, where lightning had ignited a lawn and almost burned down a house.

Later that afternoon, as we relaxed in the fire station's garage, I asked Jerry if forest fires were common. He said they were during the dry season, so I asked about the previous year. During the dry season of 2005, the western and southern portions of the Amazon rainforest experienced an intense drought, one that may have been the worst in over a hundred years.

The chief told me they'd had about twice as many forest fires as during a normal year. "We didn't have even a kiss of rain during the dry season," he said.

"How do the five of you fight forest fires?"

"The army helps us," he replied, explaining how the armed forces helped contain the fire by felling trees. He went on to describe the drought. "The river was at the lowest I've seen it. In some places you could walk across the Huallaga." I was amazed—the Huallaga, which I'd been biking along, averages over two hundred yards wide. "Some of the tributaries ran dry, and the government had to transport tanks of water to some villages. Here in Juanjuí, water was available only during certain hours. The governor declared a state of emergency."

"Did that help?"

"No. He declared it near the end of the dry season. Then the rainy season started and we had floods." He described how a number of towns along the river had flooded.

Climate scientists believe that the 2005 drought was likely due to warmer-than-normal waters in the Atlantic Ocean, affecting the timing of the rains. The warmer water in the Atlantic caused the Intertropical Convergence Zone—the ring of rain that encircles the equator and moves north and south with the seasons—to follow a more northern pattern, perhaps thus failing to bring rains to the southern and western Amazon. The warmer Atlantic was also likely responsible for the record-breaking hurricane season of 2005, which included Hurricane Katrina. But, whatever the reason, the Amazon River's flow plummeted dramatically, and forests along the southern and western edge of the basin dried and burned.

If the Amazon does transition to a savannah, the transition will be marked by many years like 2005, with fires gradually eating away at the forest. A more likely scenario, according to climate models, is that the seasonality will become stronger, with a wetter rainy season and a drier dry season.

But if rainfall does plummet, transportation could be difficult along portions of the Amazon. Although the river trip I took would likely still be possible, many locations on tributaries would no longer be accessible by large boats. During the 2005 drought, many settlements were stranded, as the river's flow was too low for ships to reach them. The Brazilian Armed Forces spent three months airlifting supplies to towns, delivering two hundred tons of food and thirty tons of medicine. Medicine was needed in part because dried-up lagoons created stagnant puddles, breeding mosquitoes and increasing the incidence of malaria.

DAY 311
September 11—Road to Uchiza

After giving a presentation at the local middle school and bidding farewell to the Juanjuí *bomberos*, I consulted with the police about road safety. The police recommended I continue south, paralleling the mountains. For a reason they couldn't explain, robberies had decreased significantly in the previous few months. They also didn't know, though, if a road connected Uchiza and Huacrachuco.

Leaving Juanjuí, I biked south on the dirt highway parallel to the two-hundred-yard-wide muddy-brown Huallaga River. Rocks on the road shook me as I progressed, and although I put more effort than normal into each pedal turn, I traveled at half the speed I would on a paved highway, and had to rest frequently to recover from the jarring ride. On one of those rests, irritated by the heat, I asked myself why I'd followed the most difficult route across the Andes. Would the Cordillera Blanca be worth all this struggle? I also carried more weight than I had in Brazil—my parents had mailed to Manaus my fourth pannier, along with warmer clothes and a thicker sleeping bag for the mountains. But those clothes were just dead weight in the jungle.

As I rode, my pedals wobbled slightly, and somewhere inside the bottom bracket—the cartridge that connects the pedal arms to the frame—I could feel that a ball bearing or a seal was damaged. If I kept riding on it, at some point the bottom bracket would seize up, and *del Fuego* would stop in his tracks. A bottom bracket can't be fixed; it must be entirely replaced. Not only might there be no road for me to ride, I might end up stuck on this side of the Andes without a functioning steed.

I entered a small village with dirt roads to look for a bike store. Because more people own bikes than cars in this region, every small town has at least one "bike mechanic." As I rode into the main square, which featured a series of one-story buildings surrounding a grass lawn, I heard a loudspeaker call someone to the town hall. "MARIA OLIVIA LA PAZ," belted out the speaker, "TELEPHONE!" Though the town had only one phone, it remarkably had an Internet café, where I could use a Windows XP machine to slowly surf the Web for fifty cents an hour. The locals looked at me with an interest I hadn't seen since the towns of Colombia.

Though the village's bike mechanic assured me he could fix *del Fuego*, when he looked at the bottom bracket he realized that the wrench and hammer in his hands would not do the job. There were many bikes in this region, but none had a Shimano bottom bracket, or its corresponding bottom-bracket tool.

I had no choice but to return to the dirt highway, whose bumps in the road masked the wobbling of my pedals. I feared, though, that the jarring was only further damaging the internal cartridge, so I tried to pedal smoothly, and used lower gears when climbing hills instead of standing on the pedals. I had no idea how long my bottom bracket would hold out—or whether there was a road ahead to hold out for. I realized how extremely lucky I had been up to that point, as *del Fuego* had throughout been as trusty a companion as I could have hoped for. A few broken spokes, a crack in the frame, or any other series of malfunctions could have by now forced a temporary glitch in my journey; I hoped we'd continue to manage so well.

After a full day of cautious biking I stopped at a clearing in the jungle where locals played soccer on a grass field with wooden poles for goals. I asked a young man on the sidelines about places to camp. He introduced himself as Lenin and, as it was nearly dinnertime, invited me to follow him to his home. He had dark short hair and wore a dirty button-down short sleeve shirt, shorts, and flip-flops.

Lenin led me along the road for a quarter mile to where two small wooden homes were tucked off the road, seemingly hiding under the thick forest. Nearby stood a wooden church about twenty feet long, with four rows of benches, a dirt floor, and no decorations. Chickens pecked at the ground near the houses. Lenin explained that he and his family are Seventh-day Adventists, a branch of evangelical Christianity, and that they watched over the community church.

Lenin's father-in-law, Yaral, told me I could spend the night by the church, and added, "In the morning, you can join us for prayer at 5 AM." I realized I had little choice in the matter. So I set up my tent by the church, cooked a pasta meal, and then returned to where the family sat on wooden benches in the space between the two wooden houses. As a warm darkness enveloped the forest, I sat and chatted with Lenin, his father- and mother-in-law, and his brothers and sisters-in-law, who ranged from about eight years old to eighteen. Candles provided modest light; they had no electricity.

"What do you do?" I asked Lenin.

"I cut down trees and sell the wood," he answered, leaning forward and describing how he uses a chain saw and selectively logs a specific type of tree, one I'd never heard of before. "That tree gets a good price on the market."

"I saw signs along the road saying IF THERE ARE NO TREES, THERE IS NO WATER," I said. "Is deforestation a problem here?" I knew that, in this region along the Andes, as much as a quarter of the forests have been cut down, mostly by farmers who had migrated over the mountains to Peru's frontier.

"Yes, but it isn't people like us," Lenin said. "It's larger companies that log huge tracts of land, and have large bulldozers. I'm just one worker with a chain saw." Though he agreed that what he was doing was illegal, he added, "I need to make a living." He suggested he wouldn't fell trees if he had other economic options, but he had a daughter to take care of.

"Yes," said Yaral, joining in. "Deforestation is a problem. If you cut down the trees, then the water disappears." They were right, of course. Not only do trees help "recycle" water by aiding the formation of clouds, but also the trees' roots prevent erosion, and create thicker soils. Since water flows more slowly off thicker soils, the forests provide a more constant source of water. I wasn't too surprised that they cut down the forests even though they knew deforestation is a problem.

"Where does your water come from?" I asked.

"We have a pipe in the side of the hill here, and it provides water for us year-round."

"Last year, during the drought, did you still have water?"

"Yes, but the flow was much less. We were lucky. Some other people struggled." He then said many fields of rice sat fallow from lack of rain.

I also chatted with Lenin's sixteen-year-old brother-in-law, Wily. "I work in a nearby farm," Wily said, "and I'm saving to try to go to college in Trujillo on the coast." He explained that he made about ten sols (three dollars) in a day of work, and that tuition would cost nine hundred sols (about three hundred dollars)—not including living costs. He said he wanted to study agricultural engineering so he could earn more through farming.

"Does coca grow here?" I asked him.

"No, that's farther up the road," he said.

"There's not much money here," said the father, Yaral, as he leaned toward me, straining to see me through the candlelight. "But we have plenty to eat." He listed the various fruits they could grow—from oranges to plantains to yucca. Then he added, "We are very lucky." I noted that what I saw as impoverished—limited employment, sharing a crowded house without electricity—he saw as fortunate. Wily and Lenin agreed with their father about their quality of life, and when I took a picture of them, they laughed at the camera and made faces, seemingly genuinely happy. Of course, what they really lacked were options—especially the ability to travel and study—but perhaps those were things Wily would yet achieve.

At five the next morning, I joined the family in a prayer led by Yaral, following a pamphlet from the Seventh-day Adventist Church. Yaral asked me if I was prepared for the second coming of Christ; I said "yes," though it probably wasn't true. He then said a prayer for my journey, and for *del Fuego*, who I'd said wasn't functioning well. And although I silently disagreed with Yaral's perspective on the Book of Revelation, I was nonetheless thankful for their prayer. Although I'm agnostic, and the scientist in me often denies the existence of higher powers interfering with our daily lives, I can't help but believe in the support of people like Yaral. I don't know what that feeling means, and people who pray and explore their spirituality would probably criticize me for not reflecting more on this connection with people and their support of my journey. All I can say is that each such interaction gave me something extra to carry with me—but instead of weighing me down, it buoyed me and strengthened me.

Before leaving, I asked if there was a road from Uchiza to Huacrachuco. "I don't know," Yaral replied, "I've never been to Uchiza." His statement surprised me, as Uchiza was only fifty miles away.

I then wrote down their address, and Lenin offered me his email address. "You use the Internet?" I asked.

"Yes, to chat," said Lenin. I chuckled to myself, but wasn't too surprised. The Internet is increasingly everywhere. And if I had to choose between having Internet or electric lighting, I'd probably choose the Internet.

Lenin, Wily, Yaral, and family

DAY 312
September 12—Coca Farms

The road led me over a low mountain ridge and then back down to the Huallaga River. Hearing the beating of a motor high above the jungle, I looked up to see, in the distance, a line of four helicopters searching for coca. I again felt nervous.

After a day of pedaling, I passed a woman selling fresh orange juice in front of a concrete house where metal rebar poked out of the roof, as if construction had halted. I stopped and paid fifty cents for some juice. She cut a few oranges, squeezed them, and handed me a glass of the sweetest pulpy liquid I think I have ever drunk. It was exactly what I needed after a day on the jarring dirt highway. I sat on the bench next to her table and ordered another glass. Soon after her husband arrived, a man named David who offered me a tour of his farm.

David told me he'd moved there a quarter century earlier with his father, in 1980, when the road was first built. His father had recently passed away, leaving David property on both sides of the road. He showed me his orange trees, lemon trees, cacao trees, yucca plants, and a small pond he had dug to raise trout.

"Do you have coca here?"

"Lots of it," he said matter-of-factly, "that's what I usually grow." He then pointed out, growing between the orange trees, small shrubs with tear-drop leaves. "I'm not taking care of these coca plants," he told me. "They're growing naturally. I used to have a big field of them, but the helicopters found the plants, and men came and pulled up the coca." David explained that on his land he could make about $1,300 a month farming coca, six times what he could make off cacao. "I have a son in college in Lima," he said. "If the eradication continues, we won't be able to afford school."

"Haven't they been eradicating for over a decade?"

"Yes, well, they did this ten years ago, but back then they sprayed herbicides from airplanes. Some of my land hasn't recovered from those chemicals. When they stopped spraying, we started growing coca again. Now we just have to wait for the helicopters to stop so we can grow more coca." He seemed to think that the campaign would end at some point.

I asked him how the previous year's drought had affected his crops.

"It was bad, but not as bad as the eradication. Actually, the coca plant is pretty resistant to drought." I almost laughed; even climate change probably won't reduce the drug trade. I then asked David about the road from Uchiza to Huacrachuco, but even though he'd lived there since the road's inception, he didn't know if it connected the two towns. I increasingly doubted the road existed—or that I'd find out before I myself got to Uchiza.

By this time it was almost 6 PM. David invited me to dinner with his wife and his two teenage daughters, and to camp in his grassy back-yard. Dinner was yucca, a stringy tuber, served with an armadillo-like animal that David had shot behind his property. The meat was fatty and juicy. I asked David what it was like living on this road.

"I remember when it was built," he said, "around 1980. European travelers used to come and explore it all the time. Then, the Shining Path took over this area for about fifteen years, and we didn't see any foreigners. The Shining Path disbanded a little more than ten years ago, but we really haven't seen foreigners down this road until very recently."

The Shining Path—a Marxist group similar in some ways to the FARC in Colombia or the FMLN (Farabundo Martí National Liberation Front) in El Salvador—staged a violent uprising during the 1980s in the hopes of imposing Communism on the country. The Shining Path controlled large swathes of rural Peru during the twenty-year conflict, during which time some seventy thousand people died. Their leader was captured in the early 1990s, after which the group went into decline. Today it's almost nonexistent.

"Why was the road so dangerous recently?" I asked. "The police told me that last January there were a few hundred attacks along this road, but that this January there were only two."

"Well, none of the people from here are thieves," David said. "When the Shining Path ruled this part of the jungle, they would kill anyone who stole anything at all—even the smallest item. So no one who grew up here would be a criminal." He explained that a group of thieves from the coast had recently moved there, and were stopping cars and robbing people. This had greatly upset David and his neighbors. "A few of my friends took matters into their own hands. Would you like some more meat?"

"You mean they got rid of the thieves?"

"Yes, they killed them. Here—these bits are especially tasty."

———

In David's retelling of the history of the region, the drought of 2005 had played only a minor role; it was the eradication of coca and the influence of the Shining Path that left a strong impression on him. He spoke of the drought only when I asked him about it.

If every year to come is like 2005, I suspect that drought would become a regular concern, a common topic of conversation. As water may be less available, and the rainy and dry seasons likely more severe, farmers will have to adapt to new varieties and new practices.

Contemplating this, I think of sixteen-year-old Wily's plan to study agricultural science so as to make a better life for himself. If global warming continues, Wily's success would be directly predicated by his ability to *afford* the training he'll need to adapt to changing farming conditions.

DAY 313
September 13—Tocache Nueva

Before I left, David gave me a bag of coca leaves, which are chewed throughout the Andes for their stimulant properties. "It's not different than drinking coffee," David told me, "and it's a hundred times weaker than cocaine." Chewing leaves has long been part of the Andean tradition.

In the midday heat, on a climb in the sun, I chewed a few coca leaves. Though my perceptions of pain and heat were dulled—and my gums were numb—my senses were heightened. It felt like I'd had a stiff cup of coffee that didn't make me anxious. I found myself worrying less about the bottom bracket, or the fact that the road ahead might not exist. These leaves could be useful.

In the slightly larger town of Tocache Nueva, I placed a long-distance call to the bike store in Palo Alto, California, that had outfitted me with gear almost a year earlier. I wanted advice on my deteriorating bottom bracket, and feared I'd have to order parts to be somehow shipped to this far side of the Andes. My call to Mike's Bikes of Palo Alto immediately put me on hold with an enthusiastic advertisement: *"Have you ever been on a ride, had your bike break down, and wish you knew how to fix it? Take Mike's Bike's on-the-road mechanics class and get yourself out of any bind!"*

When the mechanic answered his voice crackled over the poor connection. I half yelled, "I'm in Peru and need advice."

"What?"

"I'm trying to figure out what type of bottom bracket I need."

"Well, just bring it in."

"I am in Peru."

"Where's Peru?"

"South America."

"Oh, right. Oh—you're that guy who left here a year ago! How's the trip going?"

I told him my great trip was at the mercy of my bottom bracket, whose falterings I described, and that the nearest city where I might— if I was lucky—find a new one was a thousand miles away. To my relief, the mechanic said it sounded to him like I'd probably be able to ride a thousand miles before the bracket seized up, meaning most likely I would make it to Lima—but he also gave me the part numbers just in case. Though in the end he turned out to be right, the bracket continued to worsen in the meantime, so I spent much of the next few weeks afraid each next pedal would be my last.

In ending the conversation I noted how, even in this remote area, it was so easy to call Palo Alto. The connection wasn't great, but I'd been able to talk to the far side of the world.

The small town had no fire station, so I paid for a hotel room for five sols, or a dollar fifty. The manager explained that the economy had collapsed after the large operations of coca were eradicated. The hotel used to be regularly full, the rooms costing twice their current amount. That day I was one of only two guests.

I asked the manager my now holy grail question: "Do you know if there is a road from Uchiza to Huacrachuco?"

Though he didn't know, he suggested I ask Roberto, the other hotel guest, who was from Uchiza.

I finally had my answer, but Roberto confirmed what I had feared. "They're building a road," he said, "but right now it's just a trail. Back in '97, I hiked it with high school friends after graduation. It was great fun."

"I have a bike. Could I ride the trail?"

"Not really. It's very steep, and along a canyon."

My gut sank. Not only was *del Fuego* on the verge of breaking down, there was also no road even to break down on. I was stuck. I'd taken a boat up the Amazon and biked for days along this dirt highway only to be defeated by the Andes.

Deflated, I returned to my hotel room, which had a concrete floor, yellow walls, and a single incandescent bulb hanging from the ceiling. I spread out my map of South America and began estimating how long it would take to get to Lima, and then Bolivia, and then Patagonia. I wrote out a schedule. It was mid-September, and I had to make Tierra del Fuego by March in order to begin my U.S. journey with Bill. Because the road from Uchiza didn't exist, I couldn't make it to the peaks of the Cordillera Blanca. I felt completely defeated.

But then, standing over my maps in frustration, I wondered why I demanded such adventure. Despite not being able to cross the mountains from here, I could still easily finish the journey. There was plenty of time to bike farther south and then follow a more direct route to Tierra del Fuego. Who said I needed to see the highest peaks and visit every point of significance on the way?

Well, for one, a better adventure did make it easier to get attention, thus accomplishing my outreach goals. But I realized I was also driven by a desire to fully experience these places. Choosing the more difficult route had a clear reward: a more intimate, fuller understanding of each corner of the country. And I had just two main goals in Peru: I wanted to visit the capital city, Lima, and I wanted to touch the country's most remote peaks. To do so required crossing the mountains—here.

After an hour studying my map, I returned to Roberto's room to ask about other routes. "You know," he said, "you could rent a horse and hike into the mountains."

A horse! My mind quickly contemplated the logistics of disassembling *del Fuego* to be packed on the back of a horse. My heart beat faster.

"How long does it take to hike to where there are roads again?"

"About three days."

"So, I'd need a guide and a horse. How can I get those?"

I tried to wait patiently as Roberto made a call to a former teacher of his in Uchiza. Roberto told me to meet his friend the following day at noon in Uchiza's plaza. There was no official guide service in Uchiza—or organized tourism for that matter—but Roberto's friend had found someone who could help me follow the right trail.

I returned to my hotel room with newfound excitement. What had appeared to be defeat was just a new opportunity.

DAY 314
September 14—The Trail to Huacrachuco

The next day, I met my guide, Ivan, the son of a friend of the former teacher of the guy I had met at the hotel. Ivan, who was twenty years old and usually farmed coca, needed work and was eager to hike three days into the mountains for fifty dollars. He told me he was born in Huacrachuco, and had made the journey between the towns a few times.

Ivan's uncle, who lived in a wooden building where the road petered out into a trail, offered for our journey a small brown horse with skinny legs. In exchange for ninety sols, or about thirty dollars, we could rent the mare for six days—three days to hike to Huacrachuco, and three for Ivan to return.

Sitting on the porch of Ivan's uncle, I slowly disassembled *del Fuego* for the first time in a year. I unscrewed bolts, unhitched wheels, unscrewed pedals (so they wouldn't dig into the horse's back), and removed the seat, handlebars, and fender. Ivan and I hung my four panniers on the sides of the horse, and then tied the frame, wheels, and various parts on top, creating a tangle of bike parts and bags on the horse's back. It was an odd sight: all the same pieces of metal and fabric I had stared at for the past year, yet scrambled, as if my best friend suddenly had all his limbs and body parts connected at the wrong places.

The horse did not look comfortable.

"Does this horse have a name?" I asked Ivan.

Ivan chuckled. "Hembrita," he said—Little Female.

We set out, following a trail through the green canyon, paralleling a river where clear water ran over granite boulders. The moist rounded mountain ridges led away from us into clouds obscuring the distant peaks. Though we couldn't see them, the Andes have three parallel ridges between the Amazon and the Pacific; the central range was the massive Cordillera Blanca. This trail would lead to the top of the first ridge. From there, the trail connects with roads that cross a deep canyon and then lead to a 16,000-foot pass through the tallest tropical mountain range in the world. I felt myself being drawn toward the hidden peaks and the obscuring clouds.

Ivan, who had a wide frame, wore a red baseball cap and a blue T-shirt with the sleeves cut off. His square face expressed a calm but straightforward, almost simplistic, confidence. When I asked about his family he told me he had three sisters and a brother, and that they were evangelical Christians. Though most Latin Americans are Catholic, I had met a growing number of evangelicals. I asked Ivan about the difference between Evangelicals and Catholics.

Ivan paused and tilted his head. "Well, Evangelicals don't drink or dance. We also don't worship idols." Evangelicals sometimes claim that the images of Jesus and Mary are equivalent to idol worship.

Del Fuego on the back of Hembrita

"Do you drink?"

"Well, sometimes. But I don't dance."

"Do you have a girlfriend?"

He smiled. "I have four."

"Does any of them know about each other?"

"No! Of course not!"

After more discussion of his four girlfriends, I asked Ivan how he usually spent his time.

"I farm. I used to grow coca, but now, with the eradication, I work on rice fields." He told me he made about ten sols a day, about three dollars. "I make much less now because of the eradication. Of course, it's safer now. It used to be so dangerous."

"Why?"

"Well, if you said you'd sell your leaves to one man, and then sold them to another, you might get shot"—Ivan made a gun with his hand. "Life is more relaxed since I stopped selling coca leaves. Of course," he laughed, "I'm trying to plant coca right now to make money."

As we hiked I looked at the coffee plants beside the road and wondered how the landscape would be reshaped in future decades. This trail would likely become a road, and maybe many of the available slopes would be covered with agriculture. But as temperatures warm, the best zones for growing food would rise up the mountains. Coca will be grown at higher elevation, as will coffee—that is, as long as there's still reliable water. But if the region becomes drier as well as warmer, fewer crops will grow altogether, both those needed for sustenance and those grown for export.

This corridor, from the lowlands into the mountains, will likely be important for biodiversity—species will need to migrate up mountain slopes to find suitable habitats. But the rapid deforestation of the Amazon along the bases of these mountains presents a major challenge to this process. An arc of deforestation follows the road that parallels the Andes, in effect cutting off the ecosystems to which the plants and animals of the Amazon could have migrated. For this reason, many conservation strategies are promoting "corridors"—protecting land that spans a number of ecosystems so that species can move freely between them when the climate changes. Such corridors of forest still exist between the Amazon and the Andes' foothills, but if all the forests along the Huallaga River are felled, the resulting landscape might ultimately trap species that could have otherwise survived.

———

As we walked, I let Ivan listen to my iPod. He had never seen an iPod, and did not know the band U2. I switched to The Beatles. He had never heard them either, which surprised me. He had, however, seen a number of U.S. movies. He asked, "What percent of Americans know kung fu?"

"Ahh . . . not that many."

"Really? I thought a lot did."

Perhaps I traveled safely because people assumed I was an excellent fighter, or that I was carrying a gun—just like everyone in the movies.

The trail climbed along steep canyon walls. After hiking for about three hours, we shared a rest spot with two men who carried large sacks on their shoulders. The men didn't say anything to us; their limbs were skinny, their eyes tired, their skin wrinkled. They probably made this journey often—people lived along this path, and so goods were transported to them on the backs of horses or people. Continuing on the trail, we crossed a few streams, Hembrita wading through while Ivan and I jumped across rocks. Most of the hillside appeared untouched, though we encountered the occasional cultivation of cacao or coffee lining the trail. Ivan explained that we were now too high for coca, and then showed me a cacao plant, the source of chocolate, and encouraged me to suck on the sweet seed. "This land here belongs to my uncle."

"How many uncles do you have?"

"A lot."

―――――――――

That evening, we stopped in a small village, Oso, which was a handful of wooden buildings surrounding a grass field well-grazed by pigs. Ivan knew a middle-aged woman in the town who offered us shelter for the night, letting us sleep on the dirt floor in her house. When we arrived we unpacked *del Fuego* from Hembrita's back, much to the mare's relief.

Though the town had no access to electricity, the woman owned a small VCR and tiny television, which she powered on a car battery. I asked her if she'd seen other foreigners on this trail.

"Why yes. A tall group of foreign women—they were so tall and beautiful—walked through here about twenty years ago." I was shocked. I was now truly off the beaten path, in a place where foreigners rarely ventured. I felt proud and awed.

She then asked me the same questions I had heard over and over.

"Where are you from?"

"California."

She tilted her head. "Aren't there nice houses in California?"

"Ahh . . . yes. Yes there are. Why do you ask?"

"Oh, I saw them on television. They look very nice." Even though few foreigners venture here, I didn't seem surprising or otherworldly to the woman: she had seen my world on television.

I cooked a pasta meal for Ivan and myself. Later I threw my Therm-a-Rest pad on a corner of the dirt floor; Ivan slept in the opposite corner, without a mat.

We continued early the following morning on a trail that cut into the steep walls of a canyon. I watched every step of the horse as *del Fuego's* parts shifted and bounced on her back, half expecting the horror of seeing the mare lose her footing and tumble into the valley below. As the trail climbed steeper terrain the ridge lines were sharper, the drop to the river more precipitous. These steeper mountainsides were also drier, and covered with brownish green brush rather than jungle.

Following Ivan and Hembrita into the Andes

After many more hours of hiking we arrived in San Pedro, where the air was the pleasant "eternal spring" temperature of the mid-altitude tropical Andes. The town sat on the side of a ridge with a spectacular view up and down the deep valley. Most of the town's buildings were brown adobe brick, though a few were mud, even two stories tall, and a few had been plastered white. Perhaps a thousand people lived here—it was easily the largest settlement I'd visited that was accessible only by horseback or foot. I noticed that the grass courtyards in between buildings were well-grazed, and heard an occasional whinny from the horses tied to hitches nearby. Thin power lines on skinny wooden poles descended from the ridge above. "There's not much electricity here right now," Ivan said. "The water in the dam is too low. That's too bad, because it means the Internet café won't work."

"They have Internet here?"

"Yes, but it's slow."

"What do you use computers for?"

"Chat." Ivan used the Internet cafés mostly for MSN chat. While the café could cost fifty cents to a dollar per hour—a lot of money for someone making two dollars a day—that was far cheaper per minute than a long-distance phone call.

"Do you use email or read news on the Internet?" I asked.

"Huh? No. I use it to chat with my friends and girlfriends."

A relative of Ivan's let us spend the night in an extra room. After we set down our bags in the room, I took out my Allen wrenches and small travel wrench and set about the task of rebuilding *del Fuego*. We had another half day of trail ahead of us, but I figured I'd be able to bike a few portions of it, especially if Hembrita carried my bags. After spending two hours reassembling my trusty steed, it was time for dinner. Ivan led me to another building. "I'm going to get you some cuy."

For a dollar each, a woman offered to serve us cuy in her house. Her room was almost entirely dark, the town's weak electricity producing only a faint glow in the incandescent bulb above her. Four young children with dirty faces ran across the dirt floor, laughing. A puppy darted between their legs, and about fifteen small black-and-white guinea pigs squeaked and scurried beneath us. The room smelled of burned wood and moist dirt.

"Are those cuy?" I asked, looking at the guinea pigs with apprehension.

"Of course."

Great, I thought. *I'm going to eat a rat.* I had heard that guinea pigs were originally domesticated in the Andes for meat—I should have realized they were cuy.

The woman leaned over and grabbed one of the cuy, and in a few swift motions broke its neck, skinned it, and removed its guts. The children paid no attention; they had seen this many times before. The puppy, though, let out a few barks, begging for scraps; the remaining guinea pigs continued to scurry about the floor as if nothing had happened. The rodent was cooked in a pan over a wood fire in the corner.

Served with a meal of salty pasta, the guinea pig meat was greasy and gummy. I had to be careful of the small bones.

Woman who served Ivan and me cuy, and her four children

Walking out of the house, I asked Ivan, "Is that common to have cuy living on the floor of the kitchen?"

"Yes, of course, everyone in the mountains has cuy in their kitchen."

"What happens when they poop?"

"Well, you sweep the floor a few times a day."

"But that's in the kitchen!"

"So?"

We left at dawn the next day, me biking slowly or pushing *del Fuego* while Ivan either led Hembrita or rode on her back along with my bags. The trail dropped into a deep canyon—which made for some fun biking, as I bounced *del Fuego* over rocks along the skinny trail—and then the trail climbed up steeply, and Hembrita and Ivan had to wait for me several times as I pedaled with my lowest gears or just pushed the bike uphill. By midday, we'd reached the trail's end, where we encountered bulldozers and men with shovels building the road to replace the trail. I was sad to see the road being built, but I imagine the locals will appreciate being able to take a bus to other towns. If I were to return in a few years, a horse wouldn't be needed at all. Humans are changing this landscape with or without climate change—I'm sure this area will look dramatically different in fifty years.

Having now reached passable road, my journey with Ivan had come to an end. We unpacked Hembrita one last time, and Ivan left her in a pasture, telling me she'd be fine there until he returned. He then hitched a ride to Huacrachuco, where he'd visit relatives before returning to Uchiza. We agreed to meet up in Huacrachuco, after which time I'd have withdrawn the funds to pay him.

With my saddlebags again weighing me down, *del Fuego* and I followed the traffic-free dirt road to Huacrachuco, which climbed into moist cloudy meadows with grazing sheep and small homes with thick mud walls. The traffic-free road crested at a 13,000-foot pass, where rounded mountains were obscured by clouds in the thin air. Because I had ascended slowly, I had given my lungs some time to acclimatize to the lower oxygen levels of high altitudes. But, all the same, I still gasped for air as I crested the Andes.

From the pass we descended, and after an hour of bouncing down the dirt road we reached the long-sought Huacrachuco, a town of a few thousand inhabitants. Perched just below 10,000 feet in a mountain bowl, Huacrachuco was a settlement where horses outnumbered cars by about five to one, and where pedestrians, not expecting automobiles, walked down the centers of the concrete streets.

Before continuing on I reconnected with Ivan at his uncle's house, paying him and getting his contact info—which was his MSN username.

DAY 320
September 20—Marañón Cañón

From Huacrachuco, roads could lead me to the far side of the Andes, but the route was daunting: a deep canyon, and then one of the highest roads in the Western Hemisphere, cutting through the middle of the Cordillera Blanca.

I departed Huacrachuco, passing mud houses under construction and people on horses whose wide eyes told me they were unaccustomed to foreign cyclists. As the Amazon's moisture was now on the other side of the mountain crest, the hillside meadows had changed from deep green to a greenish brown. I started descending the dirt road, and then kept descending. My map showed no route here—I was only going by the advice of locals who told me this recently built road would lead me closer to the Cordillera Blanca. My altimeter showed me dropping 1,000 feet, 2,000 feet, and more; *del Fuego* kept descending. The terrain became drier, bushes lost their leaves, bare earth was exposed, and a few cacti grew up along the river. Not one car passed me, and I passed a building only once every few miles, likely the homes of shepherds. The valley walls became steeper as I dropped, the space between them claustrophobic, until finally I had lost 5,000 feet, and 60-degree slopes of rippled rock rose up from a brown river fifty feet across. I had reached the bottom of the Marañón Cañón.

I later learned that this remote canyon is one of the largest in Peru—it was as if I had accidentally biked into the bottom of the Grand Canyon. I was embarrassed that I hadn't known such a giant valley existed on my route. Now, though, I'd have to bike out of it. The Marañón felt like a giant moat, another line of defense Peru had put in my way to prevent me from reaching its far side.

The dirt road paralleled the Río Marañón for a few miles, and then crossed the river on a red bridge and started up a side canyon, headed east. The day was soon over, and I threw my sleeping bag on the ground on a gravel patch next to the river. I was awed and lonely. More important, I became tired just thinking about climbing almost 11,000 feet to reach the crest of the next mountain range.

The next day I took on the giant before me. Riding in my lowest gears, my legs spinning faster than the wheels turned, I slowly gained elevation. Higher up the sides of the canyon, arid rock gave way to greenish brown plants, followed even higher by rectangular fields. The fields, used for agriculture or grazing, looked like patches on a quilt covering the mountainside, leaving naked only the dry valley bottoms and the rock faces at the top. Though distracted by the beauty around me, I nonetheless noticed *del Fuego's* pedals wobbling more severely, and I feared for the bottom bracket.

After passing a herd of goats that briefly obstructed the road, I entered the small town of Piscobamba, where a row of two-story mud buildings with tile roofs surrounded a small plaza with well-tended bushes. A pink church with chipped paint stood at one edge of the plaza. Outside a small store sat two women, one probably in her forties, the other at least ten years older. The faces of people who live at such altitudes always appeared more wrinkled, as if the dry thin air sucked the life out more quickly. The two women wore brimmed hats made of wool, button-down wool sweaters, and long skirts. Leaning against the mud wall, they appeared shorter and squatter than they actually were. After the younger woman asked me the usual questions of where I was from and where I was going, I asked her in Spanish, "Do people speak Quechua here?" Quechua was the language of the Incas, and is still spoken throughout the Andes.

"Yes," the woman smiled in response. "We speak Spanish and Quechua."

"Can you teach me some?"

Marañón River

The woman counted to ten in Quechua, and I scribbled down the words on my note pad. Though the older woman laughed hard when I tried to speak it myself, she didn't seem to understand much of my Spanish. The first woman asked me to teach her to count to ten in English; I obliged, and I wondered if my pronunciation of Quechua sounded as bad as her English. No wonder the older woman was laughing.

I was in the realm of the former Inca Empire. At its height, in the early 1500s, it was the largest empire in the world, spanning the Andes from southern Colombia to central Chile.

The Andes seem like an unlikely place for a civilization, with towns separated by deep valleys and mountaintops that are difficult to cross even in the twenty-first century. Yet the varied terrain—the variety of climates in close proximity—promoted trade and civilization. At the highest altitudes, herders raised llamas and alpacas. In the mid altitudes, native Peruvians grew hundreds of varieties of potatoes, as well as corn and quinoa. In river valleys that reach the desert coast, natives harvested beans, squash, and cotton, and in the sea they caught an array of fish. These ecosystems were all connected by the Incas' famous roads. The road network was a system of highly improved trails built for people, not vehicles; the Incas never invented the wheel and had no draft animals. In all, twenty-five thousand miles of roads were built—a distance almost equal to the circumference of the Earth—which made accessible land one-third the size of the modern United States.

In the 1520s, smallpox brought by Spanish conquistadors killed the Peruvian emperor and a large percentage of the population. A civil war then broke out between the emperor's sons, further weakening the nation. In the 1530s, the conquistador Francisco Pizarro led an unlikely force of 168 Spaniards armed with horses, gunpowder, and steel—none of which the Incas had ever seen—and conquered the empire of several million people.

What we know about this civilization comes from oral tradition and archeology. Our knowledge is limited, as the Incas had no writing system, and the Spanish were more interested in securing gold and spreading Christianity than they were in recording history. But the conquistadors didn't eradicate indigenous culture; today millions of farmers throughout the mountains still speak native dialects of Quechua, farming mountainsides much like their ancestors had.

"Mymanta kanki," the woman said.

"What does that mean?" I asked.

"What is your name?"

"David," I said, writing for her. "Do you see other gringos biking this road?"

"Yes, all the time."

"When was the last time you saw one?"

"Hmmm . . . ," she thought. "About six months ago. They're always on bikes."

"Do all gringos pass through on bikes?"

"Oh yes, gringos are always on bikes." I smiled to myself. Given that the only foreigners who stop in this town were other cycle tourists, she must think all gringos travel by bicycle.

I left the town and continued climbing. As I encountered other people along the road, I found that saying *"Mymanta kanki"*—or any other phrase in Quechua—incited instant laughter. One man heard me and didn't reply to my question. Instead he laughed and screamed with delight in Spanish, "Look, the gringo knows Quechua!" He then served me a lunch of noodle soup and gave me a bag of salty cooked corn kernels that appeared and tasted like half-popped popcorn. "The Incas used to carry this corn in a bag and eat it on long trips," he told me.

———————

That evening, I found myself at the top of a climb, almost 13,000 feet, as a thunderstorm approached. I optimistically thought I could make it up and over the switchbacks before the storm struck. I guessed wrong. After I biked to a mere few hundred vertical feet from the summit, the heavens opened above me.

Since the mountainside lacked trees, I tried to retreat against a rock wall where the road had been cut into the steep slope. But shifting winds blew the rain horizontally, forcing water into the worn-out seams of my rain jacket. Though my gear would stay dry, as my panniers were still waterproof, I would not, and I shivered in the deluge. A flash of white light reached from the clouds to the mountain half a mile away, releasing an awesome bang that echoed as if amplified by the valley below. Water soaked through my helmet, and a disgusting sludge of sweaty sunblock dripped from the helmet pads onto my face. The rock wall might have kept me a bit safer from lighting, but I started to fear hypothermia. I jumped up and down to stay warm, fighting off cold and fear. The rain fell more heavily. Another two bolts struck within half a mile of my steel bike and me.

Grabbing *del Fuego*'s handlebars, looking down at the fender's faded image of the Virgen de Guadalupe and saying a silent prayer, I mounted the bike and descended back the way I came, noting each time the lightning struck the naked hillsides.

After riding a half-mile and nearly falling off the bike, I swerved into the first driveway I found, the only building for over a half a mile. I knocked loudly over the beating rain. The door was answered by a man with a thin mustache wearing a maroon wool blanket over his body and two wool hats.

Shivering, I asked, "I need a place to camp. Can I sleep here?"

Without showing much interest, he remarked, "You can sleep on my porch. And if you need to cook, there's a fire over there." He pointed to smoke rising from a circle of stones beneath a roof with no walls.

Shepherd who offered his porch for the night

Too cold and tired to feel appropriately lucky or thankful, I laid my sleeping pad on a mat of sheepskin beneath the porch's roof, next to an enclosure where seventy sheep huddled beneath another roof. As rain pelted the tiles above me, and thunder roared in the distance, I fell asleep.

In the morning, the clouds lifted and revealed a string of white serrations on the horizon—the sky-piercing, snow-capped glaciated peaks of the Cordillera Blanca. Once I crossed those mountains, it would be downhill all the way to Lima.

DAY 322
September 22—Cordillera Blanca

With the white mountains on my right, I climbed over the pass I had failed to cross the day before. I then descended to the town of Pomabamba, which, at maybe only half a mile wide, occupies the bottom of a high valley. On the town's edges a thin forest climbed up the valley, competing with patches of farmland and grazing land.

Most of Pomabamba's buildings were two-story adobe structures with red tile roofs, and, like Huacrachuco, almost all traffic was on foot. I noticed that some people wore baseball caps and others the traditional wool brimmed bowler hat. After a meal of pollo a la brasa at a restaurant—a common Peruvian dish of rotisseried chicken served with French fries—I headed into one of the town's Internet cafés and requested a computer from the woman running the store. Speakers in the room played traditional Peruvian folk music featuring a woman with a whining high-pitched voice accompanied by a harp. But the native song was followed by Shakira singing "Hips Don't Lie" in English.

I spent a lot of time in such cafés in random villages, often updating my website or sending emails to contacts farther down the road in plotting the remainder of my journey. These cafés made my outreach project possible, often with Windows XP machines fast enough to upload pictures as well as blog posts.

At this café, though, I noticed something new. At google.com, beneath the search bar, it said, in Spanish, "Google.com is offered in Quechua." I clicked on the link and saw two buttons beneath the search text boxes, one labeled MASK'ANA GOOGLE and the other SAMIYOQ KASANI. I was amazed.

I was again struck by the fact that, even in the most remote of villages, many people learn about and communicate with the modern world via television and the Internet. Ivan, my guide into the Andes, made only a few dollars a day, but he used MSN chat, had seen numerous U.S. movies, and assumed I knew kung fu. Peruvians on that remote trail had asked me, "Aren't there nice houses in California?"—having seen such images on television. And, perhaps most surprisingly, one can now browse the Internet in Quechua, the language of the Incas.

This new technology and flow of information may very well transform these distant villages. The cell phones and computers I saw gave me hope, since access to freely available information will undoubtedly help these economies grow. I think of Wily, the sixteen-year-old boy who planned to travel to the coast to study agriculture. If, instead of Wily crossing mountains to secure his learning, the information he seeks could be transferred to him via technology, then others like him could succeed in the same way, especially those who couldn't have afforded the travel or the schooling.

After another day of climbing, descending, and climbing again, *del Fuego* and I approached the icy peaks of the Cordillera Blanca. Clouds obscured the tops of the mountains, hiding the glaciers' sources and making it appear as if the ice descended straight from the heavens. I set my tent at 14,000 feet, about as high as any mountain in the contiguous United States. A rainstorm arrived, so I spent an entire day in the tent, acclimatizing, waiting for the weather to clear.

The next day the rain subsided and was replaced by light snow flurries. Breathing deeply to absorb as much of the thin cold air as possible, I pedaled the road upward. The route followed switchbacks through a scree field with a light dusting of snow. The narrow highway then cut through a gap in a rocky ridge that looked like a gate in castle walls.

This pass—beyond which were the blue skies of the drier western side of the Andes—would be the highest altitude I reached with *del Fuego*, and it was just as deeply satisfying as such a physical high point should be. We were surrounded by ice-covered colossal peaks, dwarfed especially by Peru's tallest mountain, Huascarán, which towers over 22,000 feet. Its rounded white cap protruded upward; below it extended a 10,000-foot-deep U-shaped valley with a thin turquoise river down its center. As I bounced down the road, gripping the handlebars with cold hands, I breathed in the light air and smiled with joy—I made it! Despite the obstacles, I'd managed to travel from the Amazon to the crest of the Andes.

Pedaling up to the Cordillera Blanca

Riding down the dirt switchbacks, feeling the air thicken, undoing days of climbing in a few hours, *del Fuego* and I dropped 8,000 feet, passing by bare rock, then by pastureland, and then by the tilted rectangles of potato fields. Although I wasn't yet out of the mountains, the route ahead would be much easier. The road led me to a town where my wheels rolled on hard surface, and I encountered the first paved highway I would bike in Peru.

PERU PART I
August 29–September 26
Presentations: 5
Flat tires: 1
Miles: 593
Trip odometer: 9,664 miles

17 PERU II 🚲
Melting Glaciers and the Desert

DAY 327
September 27—Huaráz

The city of Huaráz had something I hadn't seen in weeks: other foreign travelers. The tourists, who stood several inches taller than the Andean natives, toted backpacks and stared up wide eyed at the string of peaks whose tops radiated the white glow of snow. The travelers ambled along the sidewalk as if readjusting to walking after an eight-hour bus ride from Lima. Local residents accosted them and me with offers of hotels or guided expeditions to the mountains. After my weeks far from the gringo trail, it felt strangely refreshing to see "my own people" and be again treated like a tourist. Huaráz, a city of one hundred fifty thousand nestled in a long valley between the Cordillera Blanca and the smaller Cordillera Negra, is the tourism gateway to the mountains. Its streets were filled with white station wagons serving as taxis—rather than the horses or three-wheeled motor taxis I'd seen in surrounding regions.

After a night at the fire station I biked across town to the Huaráz office of INRENA, the government agency charged with environmental protection in Peru, to talk about the melting glaciers of the Cordillera Blanca. In addition to wanting to witness its majesty, I had visited this mountain range because its glaciers—towers of ice vital to Peru's water supply—are melting so rapidly. My objective for the remainder of my time here was to understand what it would mean for the country to lose these glaciers.

At INRENA, Oscar, a man about my age with thick jet-black hair wearing a blue sweatshirt and jeans, showed me a bulletin board with pictures. "Here are photos of the glaciers at different times," he told me as he pointed. "The glaciers in this region have all been melt-ing over the past hundred years, but over the past few decades the melting has been far more extreme." Oscar told me a quarter of the ice had melted in the past thirty years, dramatically changing the land-scape. In the bowls left vacant by the shrinking ice, there were now high-altitude lakes. In fact, as glaciers have been replaced by lakes in the previous three decades, the number of lakes in the mountains increased from just under three hundred to over five hundred.

A glacier is a moving river of ice that flows from high up on a mountain, where more snow falls in a year than can melt, to farther down the mountain, where it melts at warmer temperatures. The ice melts at a glacier's mouth, from which a constant stream of water bub-bles down the valleys. The rainfall in Huaráz is highly seasonal—Oscar told me that Huaráz may see ten inches of rain in February, but less than one inch in June and July. "The only reason water flows in the Río Santa during the dry season," he said, "is that the glaciers provide water. If we lose the glaciers, we will lose our water supply."

Oscar walked me back to where I'd left *del Fuego* and pointed out a black ridge just below the snowy fingers of the mountaintops. "You see that ridge? Snow used to fall there. It rarely does now."

I later learned that Oscar had somewhat dramatized the impor-tance of the glaciers for their water supply. Over half of the area's dry-season water may actually derive from rainwater absorbed by the soil during the rainy season. But, though it may be a more accurate assess-ment to state that losing the glaciers might decrease only one-third of the water flow, that's still an enormous challenge for a region with a growing population that's already strapped for water.

And the glaciers are at great risk. According to climate models, temperatures at altitudes over 13,000 feet will warm about twice as fast as temperatures at sea level. The reasons for greater warming at high altitudes are complicated, and concern the ways carbon dioxide affects heat flow through the atmosphere. Unfortunately, evidence has proven these predictive models to be correct: weather stations high in the Andes have warmed much faster than at sea level.

We don't know how much longer the glaciers will sit atop the Cordillera Blanca. One model suggests that by 2050, 40 to 60 percent of the remaining glaciers will be gone, and water flowing in the Río Santa during the dry season could be greatly reduced. By 2080, only a quarter of the glaciers may remain in the Cordillera Blanca, clinging to the rocky peaks of the 20,000-foot-tall mountains. In other glaciated ranges in Peru—such as the Cordillera Huayhuash or Cordillera Apolobamba, whose peaks are slightly lower—even less glacial ice will remain.

After window-shopping the town's mountaineering stores for a few hours, I finally gave in. "Want to climb mountains?" was a cry I couldn't resist, especially when expressed by Walter, the man of short stature and big voice who readily convinced me. I was an easy target: a tall unshaven gringo with a backpack, my eyes no doubt sparkling with an eagerness to explore. Of course I wanted to walk across the glaciers.

"Won't the rainy season begin soon?" I asked Walter. "Isn't it a bad time to climb? I don't have much money."

"Yes, but you can climb one of the shorter mountains and climb it quickly," he responded, pointing out Vallunaraju, a 19,000-foot snowy peak rising above Huaráz. "You're already acclimatized and can climb it in a single day." This would cost the price equivalent of $150, including rental gear and a guide. I agreed, and Walter introduced me to my guide, Wilder, who like Walter stood a foot shorter than I do. After leaving *del Fuego* at Walter's house, Wilder and I rented a taxi and drove half an hour to camp at the trailhead.

After sleeping just about two hours, we rose at one in the morning, five hours before sunrise, and first shared a cup of bitter coca tea. We then began our hike beneath a starry sky, climbing and following the cones of light from our headlamps. After two and a half hours of walking in the dark, unable to see the mountains around us, we reached the dark tongue of the glacier. I attached crampons to my rented plastic boots, my fingers struggling to secure the straps in the cold. We then paced up the ice, hearing nothing but the crunch of our crampons on the glacier.

After another hour, the mountain's silhouette emerged from the dawn's light, but the white slope gave no sense of scale. *There's the top up ahead,* I thought, *but how far?* I felt like I had lost sense of time and space during our midnight hike, and the dawn was now gradually revealing the immensity of our surroundings. Other than reaching the summit, this moment is my favorite part of mountaineering—the moment when the morning alpenglow displays the rocky ice peaks we'd climbed into during the night. Once towering over us, the long string of the Cordillera was now at eye level. In this moment I forgot the cold and exhaustion, exhilarated to be among these giants.

The light also revealed how vulnerable we were. A thin rope between Wilder and me was all that would hold me if a snow bridge gave out and I fell into a crevasse, or if my crampons failed to hold me in the snow and I slid down the steepening slope.

We reached the high ridge, where an unexpected ice wall separated us from the summit fifty feet above. "This wall isn't usually here," Wilder said, offering to help me climb around it. The routes on these mountain often change due to shifting ice, with or without climate change. I hesitated, took a deep breath, and looked at the ice face—and the 1,000-foot rock face below it. I decided it was not worth risking my life to reach the top, and told Wilder I'd come far enough.

Although the summit was just out of reach, I inhaled deeply with satisfaction. We had achieved the view we sought, a view that was worth the eight hours of climbing through dark and dawn: before us extended a line of white-peaked mountains, the tallest tropical mountain range in the world, a magnificent white serration piercing the cobalt sky.

As we descended, battling the desire to sit down and nap, the sun moved directly overhead. We had been walking for twelve hours, and now trod snow softened in the sun. Soft snow increases the chance of a wet avalanche or a collapsing snow bridge, so we didn't want to be on the glacier in the afternoon. When we reached the edge of the glacier, now seen in daylight, I noticed changes in the color of rock indicating where the ice had receded from. "Did the glacier once reach out there?" I asked Wilder, pointing about a mile beyond the current edge of the ice.

"Yes. Every time I'm up here, there's more rock exposed." He explained that on a few of the mountains they've had to change their entire routes because of melting ice. "Every time I climb the terrain is different."

The next day, back in the thicker air of Huaráz, I shared a lunch with Walter. Over a meal of salty potato soup, I asked if he was worried about the melting glaciers. "Look around you," he said. "This entire city is built on tourism. Seventy percent of the economy here is from tourists; otherwise, we have only potatoes and sheep! And our tourism is because of these mountains. What's special about these mountains? It's the glaciers. There are mountains everywhere, but only some are tall enough to have the glaciers. If we lose the ice, we will lose the tourists."

Wilder Yanac Flores

Standing on Vallunaraju, Mount Huascarán in the background

While I don't imagine the tourism industry would disappear overnight, and while I think tourists will still visit the region's tallest mountains, I agreed that something significant would be lost. The string of ice upon the mountaintops are one of nature's great works of art. When my mind pans through highlights of my journey in South America, I inevitably think back to that view from the top of Vallunaraju. Without those white caps, the mountains would lose much of their grandeur, and another natural wonder would be lost.

DAY 331
October 1—Lima

I followed the paved highway out of Huaráz, crossing a pass and the final ridge of the mountains before the long descent to the desert coast. I eagerly began, excited about the 100-mile, 14,000-foot paved descent.

It was as if someone had built a road from the top of Mount Whitney in California to sea level. The next hundred miles would be many times easier than the countless dirt roads I had willed *del Fuego* across. I would also see the Pacific for the first time in five months.

Glacier that used to reach to where this photo was taken from

From the pass a valley's deepening V led westward, following a stream that my map said would become the Fortaleza River. The air thickened, filling my lungs. As we dropped in elevation, *del Fuego* and I rode deeper into the rain shadow of the Andes, where the terrain became drier and soft spongy roadside meadows gave way to brown grass and then bare rock. It was as if I were standing still while desert enveloped the land. The Fortaleza River, though, grew as tributaries joined it, and, although it was never wider than about ten yards across, the banks became greener and more vibrant as the vegetation on the surrounding slopes vanished. Irrigation channels drew water away from the valley bottom, following contours around the valley's edge to water green pasture between the channel and the river. As I neared sea level, the valley widened into a patchwork of crops beneath the arid canyon walls. A few miles from the coast, I stopped and conversed with a group of farmers. Like most people I met in rural Peru, they were eager to talk to a strange foreigner on a bike.

Crops use mountain water to grow in the Peruvian desert

Juan, a farmer, gave me a tour of his fields, which grew chilies, corn, peppers, and a small citrus fruit. He explained how his crops were limited by the river's flow, and how the corn would be twice as tall if he had more water. He then offered me the guest room in his house. I watched *The Simpsons* with his family on their television, which was powered by a car battery they charged once a week.

The irrigation system that Juan used could probably be far more efficient. The water flowed through channels instead of through a more efficient drip irrigation system, and I guessed that his fields would yield more if he installed such efficient systems. Of course, such irrigation systems require significant investment.

The next morning found me back on *del Fuego* on a road that met the Pacific. Battling a cool headwind and thick fog, I cycled toward Lima, Peru's capital. A wide two-lane freeway with a spacious shoulder and light traffic led me south, and I covered distance at speeds impossible in the mountains. As the crow flies, 180 miles separate Juan's farm from Uchiza, where I had started crossing the mountains. That distance had required thirteen days of pedaling and hiking to traverse. From Juan's farm to Lima, a straight distance of 110 miles, I would need but two days.

Juan shows off chilies

Sand dunes lined the road, and the route offered occasional views of a rocky coastline. The desert was unlike any I'd ever seen. Ocean currents draw deep water to the surface along the coast, producing a cool thick fog similar to that of San Francisco, where I now live. The fog provided natural air conditioning, making the coast far cooler than one would expect so close to the equator. This fog also kept the air moist even though less than two inches of rain falls each year. And despite this lack of rain, every thirty miles or so the sand dunes gave way to green fields irrigated by rivers flowing from the mountains. The largest of the rivers flows off glaciers of the Cordillera Huayhuash, Peru's second-tallest range, which sits just to the south of the Cordillera Blanca. But the mountains weren't visible; the dry Andes foothills rose from the foggy ocean and blocked the view.

WELCOME TO LIMA, read the giant green sign extending above the road. As one third of Peru's 27 million people live in the capital, Lima would be the largest metropolis I had visited since Mexico City (Bogotá is slightly smaller than Lima). Tired from the long day riding into a foggy headwind, I placed a few coca leaves in my mouth. But I probably didn't need stimulants—the thrill of entering a mega-city by bike was sufficient. Of the major cities I'd entered, only Bogotá, with its bike lanes, was calming.

I passed Lima's slums on the left and right. They were filled with small brick houses partially obscured by a cool mix of fog and smog. No vegetation grew between the homes, and instead there was just the brown dirt of the desert. Trash lined the edge of the road, and a few feral dogs searched the debris for nourishment. My head turned on a swivel, watching traffic, and I tried to keep a buffer between me and the thundering buses and trucks that blew diesel fumes into the air.

Considering myself in relation to these slums, I realized how artificial, in some ways, my experience of Peru was. I had spent most of my time in the countryside, yet the majority of Peruvians now live in cities, many in slums like those I merely zipped through. The growth of Lima and other cities in Latin America in the last half century has been remarkable. In 1950 less than 40 percent of Latin Americans lived in cities; today that figure is almost 80 percent.

One drawback of bicycle travel is that I don't interact much with the urban poor. While I stayed with poor campesinos in the countryside, in the cities I slept either at fire stations or at the houses of well-off individuals I'd contacted via email. I never experienced the homes and lifestyle of a large portion of the population.

In some ways, the urban poor have a higher quality of life than their rural counterparts. The infant mortality rate in the city is half that of the countryside, and incomes are higher. Basic services are better. In the countryside three-quarters of households don't have water piped to their house; in the city less than a third lack this service. Less than one-twentieth of houses in the countryside have a modern toilet; two-thirds have such luxury in the city. In the countryside, nearly one-third of the households have more than five residents per room; in the city, that figure is only about one-tenth.

But as I looked at the slums, I wasn't sure I'd choose to live there instead of the countryside. While there's more opportunity for financial success in the city, there's also more opportunity for failure. If one can't get a job in the countryside, one can still grow potatoes, maize, or beans. Also, not having a toilet or running water is less of a health hazard in the countryside than it is in the city.

Slums on the outskirts of Lima

When people travel from the country to the city looking for work, they generally end up in the slums. During times of civil unrest, slums can grow faster than does their ability to absorb the influx. In the 1980s, a period of economic stagnation and the terrorism of the Shining Path, the city of Lima annually gained as many as four hundred thousand rural villagers trying to escape violence and poverty. Similar bursts in migration have occurred throughout Latin America during times of economic hardship and rural violence.

As I rode toward the city center I noticed the buildings along the road were larger, and made of clay bricks rather than mud bricks. Many of the buildings had steel rebar sticking out from their flat roofs, as if only half finished. I followed a ramp onto the road Universitaria, a wide boulevard that remarkably had a bike path blocked from traffic by an asphalt curb—though I frequently had to veer into traffic to avoid street vendors parked in the lane. I then crossed the Río Rímac, a small trickle of black sludge maybe twenty feet across. The river passed through banks so thick with garbage that the river appeared to carve its path by eroding layers of trash.

But inside the ring of slums I encountered a wealthier and more commercial district, where I passed a Pesca gas station, a Blockbuster Video, a McDonald's, and an ACE Home Center. Navigating with my thirty-page map of Lima, I rode through the financial district San Isidro, which had taller semi-modern buildings. The traffic was horrible, buses and cars jockeying for position. Lima had neither a metro nor an efficient bus system like Bogotá. (They have since built a Bogotá-style bus system, but it has only one line.) I was faster on my bike than any car, but only because I didn't fear riding through the red lights or the gaps between cars stopped in traffic.

In Miraflores, the district where the middle class and wealthy live, I stayed with José, a colleague of the reporter who had hosted me in Mexico City. By this point in my journey I was no longer surprised at how people helped me find lodging, schools to visit, or publicity outlets. José, an editor for *El Peruano*, one of Peru's largest newspapers, did all three: he not only offered me room and board, but also wrote an article about Ride for Climate and helped set up a talk at a school where his sister teaches. However unsurprising this help was, it still felt magical. I feel so grateful that I found so many people eager to promote my project, and that I was able to share the journey with my hosts, the media, and classrooms of students.

While in Lima, after a few days at José's home, I biked a short distance across town to the Consejo Nacional del Ambiente (National Council of the Environment), or CONAM. A woman working there talked with me and shared their publications on global warming.

"A hydrologist just studied our water supply, which comes from the Río Rimac," she told me.

"You get your water from that filthy stream?" I asked, unable to hide my surprise.

"Far upstream of the city, of course. And the river would be much larger if we weren't taking water from it." She told me that, during the dry season, a significant portion of their water derived from glacial melt. That water also powers their hydroelectric plants. In fact, 80 percent of the country's electricity comes from hydroelectric power; during the dry season, most of of the water to power those dams comes from glacial water. The glaciers that are the source of the Río Rímac sit on mountains that aren't as tall as the Cordillera Blanca, and consequently

Lima traffic

the glaciers are likely to entirely disappear much sooner. "When the glaciers are gone," she said, "we'll have to figure out new sources of water and energy."

"Do you know how much that will cost Peru?"

"I'm not sure. The World Bank is studying it. The cost will be in the billions of dollars."

"Where will Lima get its electricity when the hydroelectric dams have no water?"

"We have a lot of coal in the mountains. I think we'll burn coal."

———

She was probably right. Peru's use of coal is already increasing rapidly; in the past decade Peru has doubled the amount of coal it burns. The average Peruvian currently uses one-fifteenth the electricity the average U.S. citizen uses. If the Peruvian economy grows, the country's use of electricity can only rise.

Likewise, water use is likely to increase. Nearly 30 percent of houses in Lima are not connected to running water. Considering that I saw almost no gardens in the city's slums, and the fact that a large percentage of homes don't have running water, it's unlikely that conservation will save much water—they're using so little to start with. As the

Article about Ride for Climate in *El Peruano*

city's population continues to grow, and as its economy grows, so will demand for water.

Lima is not alone in relying on glacial melt for its water and electricity. One-sixth of the world's population gets water from rivers fed by snow and glaciers. From the Colorado River in the U.S., to the Rhine in Europe, to the Yangtze in China, to the Indus in Pakistan, the seasonal flow of water of countless rivers is partially determined by the melting of snow and ice in the mountains. In China and India in particular, hundreds of millions of people rely on meltwater from Himalayan glaciers.

In general, the regions that have snow and ice—higher latitudes and higher altitudes—will warm more than will other regions on the planet. Also, winter temperatures will likely warm more than will summer temperatures, which is bad for snowpack and glaciers. In just a few decades, rivers around the world will be far more variable, and many rivers, including those in Peru, will see less flow during the dry months, when water is needed most.

The variability in water supply might be overcome by engineering projects such as more and larger dams, or perhaps desalination plants, both of which would be expensive. Without major investment, the farmer Juan, with whom I stayed on the coast just a few days before, would have to adjust to lower yields or abandon his farm altogether. Maybe he'd move to the slums of Lima to look for work. In the extreme, climate

change could trigger massive migrations from the countryside to the cities, just as has occurred at various times throughout history. But while, in
one example, Lima swelled in the 1980s on account of economic collapse
and social unrest, should major climate change come to pass, the migrations would likely be many around the globe, possibly simultaneously.

DAY 343
October 13—Machu Picchu

To my great relief, one of the few high-end bike stores in the capital had a
replacement for *del Fuego's* wobbly bottom bracket. After basic repairs and
an overhaul of the bike's hubs, we departed Lima and rode south.

Camping by llamas high in the Andes

All throughout my journey, when people asked me what my
favorite country or town was, I always answered saying *their* country or *their* town was my favorite. Though they rarely believed me,
I at least got a laugh. Truthfully, though, once I'd covered my entire
itinerary and *del Fuego* had rolled on his last stretch of Latin America—ultimately my favorite country was Peru. Part of the reason was
the topography: the Andes provided breathtaking landscapes. But it's
actually the humble dirt roads that really stayed with me—in no other

country did I experience similarly remote mountain roads. And the people along these dirt roads were almost always more friendly to me than were those I encountered along the paved ones. When I daydream about my bicycle adventure, my mind often drifts back to those lonely roads.

After two days on the coast, I followed a road back into the mountains and traveled for two weeks south on dirt roads. Six-thousand-foot climbs, followed by bouncy descents, rewarded *del Fuego* and me with spectacular vistas. I met more locals who thought that all gringos travel by bicycle, picked up more phrases in Quechua, and stirred more exclamations of "Look, the gringo knows Quechua! Ha ha ha!" High in the mountains, I saw herds of llamas and alpacas, animals domesticated by the native people for their fur and meat. Willing *del Fuego* across mountain passes was a daily challenge; I wasn't surprised that the Incas never invented the wheel.

Del Fuego on a road high in the Andes

I spent November fifth, the one-year anniversary of the start of my journey, at the ruins of Machu Picchu, the famous Incan city that was never found or plundered by the Spaniards. The ruins occupy a ridge 1,000 feet above a wide meander of the Urubamba River. The city is the tourist Mecca of South America, providing the perfect combination of archeological significance and scenic beauty, and many foreign tourists walked between the five-hundred-year-old buildings, digital cameras in hand. On the eastern wetter side of the Andes, Machu Picchu's grass radiated green, and the rounded mountainsides were covered with thick vegetation.

As I watched fingers of clouds intermittently reveal the ruins, I felt intense gratitude for my year on the road. I thought back to what John Guiliano had said in El Salvador about incurring debt to the world while traveling by bicycle. I was often lonely and tired on my journey, but those emotions were always dwarfed by the gratitude I felt for the goodwill I received from the many people I encountered along the way. In town after town after town, the people of the lands I traveled through continually offered me food, lodging, and safe travel. It was as if everyone I met contributed to a mutually expressed statement: "The road to Tierra del Fuego is generous."

Peruvian family whose house I camped next to

Machu Picchu

PERU PART II
September 27–November 14
Presentations: 4
Flat tires: 1
Miles: 1,371
Trip odometer: 11,035 miles

Bolivia route

18 BOLIVIA
High and Dry

Bolivia population: 10 million
Annual per capita GDP: $5,300
Annual per capita CO$_2$ emissions from fossil fuels: 1.5 tons

DAY 376
November 15—La Paz

La Paz, Bolivia, the highest elevation capital city in the world, may also be one of its most scenic, as it fills a spectacular dry canyon between a high plateau and the glaciated peaks of the Cordillera Real. A small handful of twenty-story buildings rise up in the canyon's center, and brick structures climb up the valley's edges until the walls become too steep, revealing bare rock. Beyond La Paz, opposite the plateau, the canyon's walls lead to ridges that connect to the four-mile-tall mountain Illimani, which resembles a giant glaciated pyramid, a monument over the city.

I spent a week in the wealthier "low-elevation" district of the valley, which was a mere 10,000 feet above sea level instead of 12,000 or 13,000, and where daily highs hovered in the 60s Fahrenheit instead of the low 50s. I stayed with two American teachers who worked at the English-speaking private school. I spoke to students at the school—most of whom were wealthy foreigners—and then spent Thanksgiving with a group of U.S. citizens in a comfortable three-story house at the base of the valley. Though Bolivia is one of the poorest nations in the

Western Hemisphere, I ate turkey, mashed potatoes, and enjoyed ample servings of wine, momentarily forgetting I was more than a year away from home. It felt wonderful but also somehow improperly indulgent. Of course, I also wondered if it is only proximity that makes such disparities in wealth feel inappropriate.

On my last day in La Paz I woke early, as a contact from the school had arranged an interview on *Al Despertar,* Bolivia's equivalent of *Good Morning America.* After biking up a hill and wheeling *del Fuego* into the television studios—which occupied a fairly plain brick building—I was escorted to the room where bright lights illuminated the set, which included a yellow couch and coffee table in front of bamboo and a painted forest scene.

Waiting my turn and setting up *del Fuego* on the far side of the stage, I watched the interview in progress. The guest was Bolivia's Secretary of Labor. He spoke quietly and had the dark skin and facial features of Andean natives. I couldn't quite make out what he said—he seemed to mumble his words—but I was struck by the contrast between him and the newscasters, who were both tall with fair skin, the woman with dark blonde hair. The newscasters, as with television personalities across the continent, were far more European than indigenous. Ever since the Spaniards conquered the region hundreds of years ago, installing a semi-feudal government, those of European descent have served in government and owned the land, while the indigenous people formed the underclass. I remember private school students in Mexico who told me they use the word *indígena* as an insult.

Bolivia may be overcoming this bias, as even their president, Evo Morales, is of indigenous descent. Morales grew up speaking a native language, Aymara, and herding his family's llamas. He later became the leader of the coca-growing union, fighting for rights of coca farmers. He also argued for indigenous rights—and used anti-American, anti-capitalism rhetoric—which only increased his popularity. He gradually rose in politics and was elected the nation's president in 2006, becoming the first fully indigenous head of state in Bolivia since the Spanish conquered the region nearly five centuries earlier.

Evo Morales has definitely sparked controversy, as he leads a political party called Movement Toward Socialism and was a close ally of Hugo Chávez. Bolivia itself is deeply divided; the lowland regions to the east—where the country's rich natural gas resources are located—are wealthier, ethnically more European, and fiercely opposed to Evo. Evo, like Chávez, has attempted to rewrite the constitution and redistribute the natural gas wealth. He has also politically divided the wealthy and the poor in Bolivia much as Chávez has in Venezuela.

I looked again at the newscasters, noting how I resembled them more than I did their guest. This made me wonder: would it have been more difficult to publicize my project throughout Latin America if I were shorter and had darker skin? In riding across this region, I certainly couldn't hide my European ancestry. And I knew only what it's like to travel as myself, looking like a tall male gringo. I wondered, was it easier to travel because I looked like someone on television? No, scratch that; I didn't *look* like someone on television—I *was* someone on television.

"And next," the host told the audience, "we have a scientist who has biked all the way from California to talk about a serious environmental problem." They walked over to *del Fuego* and me as the camera panned to my side of the stage.

The co-host asked the standard questions—how long have I been on the road, where do I sleep, how far do I bike each day. I showed off the contents of my faded panniers. She then asked me about climate change, and I explained how burning fossil fuels creates carbon dioxide, how we've increased the amount of carbon dioxide in the atmosphere by more than a third, and how this could lead to a much warmer planet.

"And, here in Bolivia, what will be the consequences?"

"Well, in the mountains over La Paz, the glaciers that feed your rivers are melting. The glaciers provide the city with drinking water and also power hydroelectric dams above the city. When we lose those glaciers, there may be water shortages." (The glaciers above the city are melting rapidly. The city actually has a high-altitude ski resort, but snow no longer reaches the bottom of the resort's only lift.) "Also, many Bolivians are farmers; global warming, through more storms and droughts, will make it more difficult to survive in the countryside."

"Well, what should we do?"

"We need to do three things: make clean energy cheaper, use energy more efficiently, and slow down deforestation. To do this, we all need to work together." I added that, once I reached Tierra del Fuego, I'd cross the United States as well, to try to get my own people to act on this issue. It felt good to thus feel more authentic.

The co-host then said, "Well, thank you so much for joining us this morning. We would like to give you something from one of our sponsors." She gave me a twenty-ounce bottle of Coke. "Here is a refreshing drink for your ride. Enjoy Coca-Cola." The two hosts then each opened their own Coke bottles to enjoy.

The next guests were identical twin brothers who had married identical twin sisters, for whom the host brought out small identical chocolate wedding cakes to celebrate. A famous Bolivian pop singer, who had long curly hair and used crutches on account of his early-age polio, hobbled onstage and serenaded the couples. As the show seemed to have progressed from the serious to the less serious, I guessed I had somehow been the transition.

Morning TV show *Al Despertar*

DAY 384
November 23—The Altiplano

After the interview I met up with Wouter, a skinny Belgian cyclist my age I'd met on the bus from Machu Picchu to Cusco. I had been telling

Wouter how sore my legs were from hiking around the ruins all day, and how that was strange since I'd spent the past year biking from California. Wouter had replied he also was sore, even though *he'd* spent the past year biking from Mexico. There had been a moment of silence as we realized we'd randomly sat next to someone else who was biking across the continent. Two days later, we'd joined forces and pedaled out of Cusco, deciding to cross Bolivia together—reconnecting after my week of outreach in La Paz. As it had been nearly three months since I'd biked with a companion, I welcomed the company.

We biked upward out of La Paz's deep canyon, having decided to follow the drier western edge of the landlocked country south, along the high plateau known as the Altiplano. When we reached the high, cold plateau, we donned our arm and leg warmers and gloves to shield us from the 45-degree air. Above La Paz, at over 13,000 feet, sprawled the city of El Alto, a series of two-story brick buildings standing in the cold thin air. Small Volkswagen buses jockeyed for position on the road. Though there were few billboards, the side of one brick building was painted with a tall advertisement for Coka Quina, a Bolivian cola. Crowding the sidewalk were men wearing outfits free of nationality—baseball caps, jeans, and sneakers. The women, however, seemed to follow a more Bolivian dress code. Many wore a bowler hat, made of thickly woven wool, that looked like a cross between a top hat and a cowboy hat. They also wore dresses that poofed out, hiding their curves. Many wore colorful shawls around their necks, their hair in two dark braids.

Continuing south, we exited the city and began riding through the wide-open plains of the Andean Altiplano. The two-hundred-mile-wide, two-mile-high plateau deeply disoriented me—the terrain was unlike anything I'd seen in my year of cycling. Only a few unimpressive ridges punctuated the plateau. As there were few trees, we had a clear view across a terrain covered with only short green shrubs. The low-relief landscape suggested we weren't at high altitude, but the thin air, deep blue sky, and strong sun told otherwise. Were we at sea level, or high in the clouds?

"Hey, Wouter, slow down!" I yelled ahead. Wouter rode a mountain bike without shocks, much like *del Fuego*, but instead of using panniers he pulled a trailer, and was always far out ahead.

"Come on—you Americans are so lazy," he yelled back, joking.

"Cut me some slack; I'm not chewing coca."

Wouter smiled back at me with a green mouth. Chewing coca leaves is common in Bolivia—we often passed villagers who gave us green smiles—and Wouter had taken the opportunity of indulging heavily in the local custom.

Wouter—who is fluent in five languages, including English and Spanish—used to work as a bartender in Belgium. Nearly a year earlier, feeling somewhat depressed and clueless about what he was doing in life, he quit his job and flew to Mexico City with a few thousand euros in savings. He then purchased a bike and started riding south—though he wasn't pedaling to save the environment, cure cancer, or help impoverished children. "You Americans," he said, "are all riding for causes. Why can't you just ride?"

I enjoyed Wouter's company. He had a dry wit and an enthusiasm for riding, and for the first time I could talk about my past year of biking with someone who'd similarly experienced the same roads. As we biked, we discussed the army checkpoints in Colombia, the quality of tacos in Mexico, the beauty of the Cordillera Blanca, and the challenge of Central America's crime-ridden cities. Wouter told me of the various other cyclists he'd ridden with: two American brothers biking to raise money to fight brain cancer, a German cyclist who biked the entire route in blue jeans, and my friend Gregg, who'd continued riding after Brooks returned home. "But I lost Gregg when I spent a month in Peru," Wouter said. He'd met a Peruvian girl in Huaráz and spent a few extra weeks in the city. (They have since married and live in Belgium.)

Near dusk, after a day in the thin air of the disorienting Altiplano, Wouter and I rode into a tiny village. It was just a row of mud buildings along a dirt path, its few isolated twenty-foot-tall trees contrasting with the otherwise tree-free horizon.

A wrinkled man wearing a thick red dust-stained jacket offered to let us sleep in the school where he taught, a tiny one-room adobe building painted white and blue. When he unlocked the door we saw a classroom barely large enough for our two bikes and sleeping bags.

"What do people grow here?" I asked him.

"Potatoes and quinoa. Quinoa grows very well here."

I told him how I'd been giving talks at schools, to which he replied the schools had just started vacation. Of course, as was the case at many such villages, I wasn't sure what I'd say to the children, who probably dreamed of lives that happened to use more energy rather than less. Twenty-five percent of Bolivians live on less than two dollars a day, and a similar percentage is malnourished.

Wouter Tas

Residents of a small village in Bolivia

The school appeared to have pursued an energy-use reduction program, but this was mostly because its lightbulbs were burned out. For fifty cents each, Wouter and I purchased new bulbs for our "guest house" at the town's general store (though we were surprised that the store—a one-room mud shack attached to someone's house—even sold lightbulbs). In the morning, as we cooked breakfast and prepared our bikes outside the school, a group of children, wearing brimmed hats for the sun and jackets to keep warm, surrounded us with a quiet curiosity. Like many children I met in the countryside, they were too curious to drop their gaze, but they were also too afraid to ask a question. I talked only with the teacher.

Wouter and I continued south along the highway, which paralleled a small lake and swampland. During the past ice age, this region was wetter, and much of the Altiplano was covered by a series of lakes, the largest of which, Lake Minchin, was the size of modern-day Lake Michigan. These lakes had sediments that smoothed the land, and when a drier climate evaporated the water, this plateau was left behind. A few remnants of the ancient lake remain. Titicaca, South America's largest lake, straddles the Peru-Bolivia border. Titicaca empties into the Desaguadero River, which flows south and ends at a salt marsh, where the remaining water evaporates. Even farther south across the Altiplano, the terrain becomes more arid, the landscape dotted with salt flats, all of which sit in basins with no outlet to the sea.

As these basins have no rivers flowing out of them, the balance of evaporation and precipitation determines the levels of the lakes. But during the last ice age, more rain fell, and less moisture evaporated in the colder temperatures. As a result the basins filled with water, creating a five-hundred-foot-deep lake where we now rode. Global warming will almost certainly change both evaporation and rainfall in this region, likely again altering lake levels.

Many other parts of the world have similar lakes. The Great Salt Lake in Utah rises and falls depending on rain levels, such that a rise in the lake during the 1980s threatened to inundate Salt Lake City's airport. The Caspian Sea, the world's largest lake, has no outlets, so the balance of evaporation and rain determines its levels. In the past century, during comparatively mild climate change, its level fluctuated by twenty feet—which, in the past few decades, flooded oil rigs along its northern coast, creating an ecological disaster. If the Earth's rain patterns change as we predict they might, these basins would be dramatically altered.

According to climate models, Bolivia is likely to become drier rather than wetter—a prediction made with the caveat that climate in the region is difficult to predict, partially because of El Niño. El Niño is the result of a huge oscillation in the equatorial Pacific that affects weather across the globe. The "normal" state of the equatorial Pacific is that winds blow from east to west. Because storms move from east to west, the west coast of central South America is in the rain shadow of the Andes, and thus is extremely dry. The westward-blowing winds also pull water away from the coast, and draw cold deep water to the surface. Consequently, this coast is shrouded in a cool fog.

During an El Niño year, the surface winds weaken, and the water near the coast is thus warmer than usual. These warmer waters slightly reverse the normal weather patterns, making the coast wetter than normal and bringing drought to the Amazon and Altiplano. But when the winds blow more strongly than normal, during the opposite of El Niño—La Niña—the situation reverses. In 2000, during a La Niña year, heavy rain caused Lake Titicaca to rise by five feet, and two hundred thousand people found their homes under water.

Scientists debate how El Niño will respond to climate change. Some models predict a stronger El Niño, others a stronger La Niña. Earth's history, though, suggests a permanent El Niño might be more likely. During the Pliocene Epoch of about three million years ago, the Earth was about 5 degrees Fahrenheit warmer than it is today. According to some, but not all, paleoclimatologists, the Pacific was in a permanent El Niño state during this period.

Though we aren't sure how rainfall will change in the tropics, we are doing a global experiment to find out what will happen. In a few decades we'll learn if settlements in the Altiplano need to relocate from the shores of rising lakes, or if they need to adjust to diminishing rainfall.

DAY 386
November 25—Salar de Uyuni

Wouter and I passed through several more villages of adobe houses, all with residents who stared at us quietly. Herds of llamas and alpacas casually crossing the road scurried off when our rumbling bikes approached. The terrain became browner and drier, and the road turned to dirt, slowing our progress. But I met the dirt roads with excitement. We were nearing the remote corner of Bolivia, where its famous salt flats are found. Much as I had longed to see the Cordillera Blanca in Peru, I was eager to cross the famed Salar de Uyuni.

Wouter was developing a rash on his rear—saddle sores—from putting in long days. I told him he should rest, but he opted instead to just chew more coca leaves, after which he picked up the pace. As a result we made it to a small town for the night, paying a dollar and a half each for a hotel room.

The next day we continued into terrain that became increasingly otherworldly. We biked a sandy road around the edge of a red volcano, and then gained a view of a massive white surface spreading to the horizon: the Salar de Uyuni, the world's largest salt flat—a sheet of salt half the size of New Jersey. Beneath a deep blue sky the surface shone like ice; sixty miles away, rising from the horizon, the outline of brown volcanoes marked the salar's far side. A number of trans-continental cyclists had told me the Salar de Uyuni was one of the highlights of their entire journey. It is perhaps the only place where you can leave roads entirely and yet still ride. I gripped the handlebars and smiled—I couldn't wait to bike across it.

Once we reached the surface we pedaled out onto the salt. Wouter and I yelled with delight. I took out my speakers, which I'd rarely used, plugged in my iPod, and played songs from my "pump-up mix"—a few tracks from U2, the theme from *2001: A Space Odyssey*, and Queen's "We Are the Champions." We rode around in circles, amazed that we could bike in any direction we chose. After some frolic we charted a course toward a notch in the mountains, as we were told in that direction we'd find an island in the middle of the salt flat.

In the rainy season, a thin film of water accumulates across the plain; during the dry months the water evaporates, leaving yet another layer of salt behind. Even though the flat stands at 12,000 feet, an elevation at which one usually finds steep mountain slopes, the cyclic flooding and evaporation has left the Salar de Uyuni the flattest region on Earth. Its vast surface varies by less than two feet in elevation; that's flat enough for satellites to use its surface to calibrate their altitude measurements. When water evaporates from the flat, it leaves behind tiny white ridges that form a pattern of polygons, each a few feet across, linked up like a giant puzzle. The ridges crunched under our tires.

"Hey, Wouter—I'm going to close my eyes and bike. You tell me to steer right and left to stay on course."

"You trust me?"

"What could I run into here?"

I closed my eyes and pedaled. As I have never biked with my eyes closed before, I had to force myself to keep them tightly shut. I remember feeling how the dry cool air flowed by my skin, how the strong sun radiated above, how the handlebars shook from the rough surface. I heard only the cracking of salt and Wouter's voice. "A little right," he'd say. "A little left." *There's no fear of crashing into anything, right?* I became nervous, feeling claustrophobic, and also feared I'd hit some obstacle and fly over the handlebars. I finally opened my eyes. But, strangely, I wondered if I had ever closed them. The salt still stretched to the cool horizon, and claustrophobia was replaced by its opposite: the awe of too much space.

Wouter on the Salar de Uyuni. We stayed on the tall island in the middle.

We had enough water for two days, and could in theory stay any-where for the night. But somehow it seemed inappropriate to sleep on the vast white surface, and part of me feared it was actually ice, and that we'd somehow fall through it. I was glad to be riding across the salar with another person, as I also feared getting lost. A companion would also help me from going crazy on the unending surface.

Gradually, the distant small peaks rose from the salt, "islands" in the flat's center. The largest was perhaps a mile and a half long and a few hundred feet tall. After biking around its edge, we found its only inhabitants were large rocks, tiny bushes, and cacti with small furry needles.

We set up our tents on the "shore." As the sun dropped low in the sky, Wouter and I hiked up the rocky island, careful to avoid the cacti.

We reached the top just as the sun graced the mountains on the rim of the salar, earning us a striking vista that easily ranked among the most spectacular views of my entire trip. A few isolated islands poked up in the distance, black mounds protruding from the white surface. The glow of the setting sun reflected on the salt, contrasting with the dark outlines of peaks along the curved horizon. Turning around, we saw the rim of mountains surrounding us, as if we were at the center of a giant amphitheater—from which, atop that island, we could address the entirety of South America. My heart quickened, awed by the view.

"How does that feel, South America? Huh? Yeah, how does that feel! I'm biking across you!" I yelled as if I could taunt a whole continent, laughing with amazement at where we were, and how we'd arrived. Unlike at other places, where I was filled with gratefulness and wonder, here for some reason I felt like gloating. And it felt great. Maybe it was because there was no one else as far as we could see, and we felt like the kings of this empty land. Or maybe the salt was making me delusional, believing I had come to conquer the continent rather than to share a message of cooperation. "I biked here from California!" I laughed.

Wouter laughed too, and shouted in agreement. "Yeah! How does that feel?!"

We yelled again as loud as we could, drunk on our accomplishment; only the salt and the cacti heard us.

As the glow faded behind the rim of mountains, Wouter turned to me. "Dave, did you remember your flashlight?"

"Ah, no, I left it at camp. Did you?"

"No."

"Oh. That's interesting."

We stumbled down the boulders, this time *very* attentively avoiding the cacti. The waxing Moon provided just enough light to illuminate the ground's contours. Awed by our surroundings, I placed my feet carefully.

———

Much later, I learned that the salt flats of Bolivia and northern Argentina hold a valuable resource for the "green" economy: the element lithium. It's used to make lithium-ion batteries, the battery of choice for electric vehicles, and half of the world's known supplies are in the salt flats of this region. The metal is mined by shoveling the salt into a series of evaporative ponds.

Currently, the only practical way to eliminate the carbon dioxide emissions from transportation is to use vehicles run by renewably charged batteries. To do so on a global scale will require immense amounts of lithium. Today, there are one billion cars in the world, and it is estimated that in the next few decades human society will add another billion. According to some estimates, there's enough lithium in just Bolivia and northern Argentina and Chile to provide batteries for more than four billion cars. That means there will be enough of the element should we decide to power the world's vehicles on batteries.

But extracting a quarter or more of this region's lithium would greatly alter the wilderness we biked across. The surface of the salt flat would be scraped into piles and then processed in evaporative ponds, ultimately covering the region. (Note it is unclear where such an operation would get sufficient water.) The dirt roads leading to this remote corner of the country would probably be replaced by paved highways to deliver workers and transport goods. While all this would somewhat disrupt the surrounding beauty, I still believe it would be worth the end result: replacing petroleum. Perhaps we could strike a balance, setting aside some of the most remarkable or ecologically important salt flats for conservation. Plus, given that there is plenty lithium to be found across the planet, we need not mine the entire Uyuni salt flat. Also, it should be noted that many forms of oil extraction also incur great—or much greater—ecological harm. In any case, such are some of the decisions we'll have to make about how we manage this planet.

The following day we found ourselves rejoining humanity when we explored another set of islands—which also attracted many dozens of tourists traveling by jeep. They rolled down their windows, leaned out, and took pictures of us as if we were gazelles on the Serengeti. We retaliated with our own style of photograph.

That evening, Wouter and I split ways. He had lost his bank card somewhere along our route, and the rash on his rear had become so bad no amount of coca leaves could dull the pain. So Wouter would hitch a ride with tourists back to the city of Oruro. As usual, though I'd been glad for a companion to add variety and perspective to the journey, I was also excited to travel alone again.

I planned to ride quickly south across uninhabited wastelands. The Moon would be full in a few days, allowing me to bike night and day across the Atacama Desert. I was eager to put on miles now to avoid being rushed at the end of my journey. I also longed to find my physical limit, to see how far and how long I could ride. Despite having biked for more than a year, I hadn't yet spent a month without a long break, or many short breaks, so I wanted to see what would happen if I just pushed ahead and biked. Perhaps I also feared becoming bored or losing my purpose before I reached the end of the road.

Wouter Tas

Fun with perspective on the salt flat

Following a jeep-worn path across the Salar de Uyuni at dusk

Whatever my motivation, as dusk fell I cycled south, following a jeep path worn in the salt. The sun set behind the ring of mountains to my right, illumination soon to be replaced by the waxing Moon to my left. Wouter later told me he had watched me ride away from the top of the island, watched me get smaller and smaller as I approached the horizon. Then I disappeared across the salt, becoming a tiny shadow reaching toward the horizon.

BOLIVIA

November 15–30
Presentations: 2
Flat tires: 0
Miles: 547
Trip odometer: 11,582 miles

PART V
CHILE AND ARGENTINA 🚲
Wealth & Climate

═══════════════════════════════════════

Perhaps the most lasting impression from my journey, apart from the epic scenery, was simply witnessing the differences in living standards—the fact that only some people in the world are rich, while a great many are poor. That vast discrepancy leaves most of us feeling rather sick to the stomach, and also a bit confused. Is it fair that I can (relatively) easily find a job earning perhaps as much as a hundred times what many people in poorer nations earn? And if it isn't fair, what is my moral obligation to do something about it? Simply donating money doesn't necessarily help—it doesn't "teach someone to fish." And development economists, the people who study how to help poor countries improve their lifestyles, often don't agree on what we should do to help—leaving those of us who want to help, but who aren't experts, confused about what to do. But I think for the "haves" to do nothing for the "have-nots" is a cop-out, and is perhaps one of the worst responses. We share this planet with more than seven billion people. Not caring about the inequality of *how* we share the world just feels wrong.

As I traveled from Peru to the tip of the continent, my thoughts on climate change and poverty were still evolving, and I would soon experience countries that seemed to be winning economically. The poverty I'd seen in Peru and Bolivia provided a startling contrast to the emerging wealth of Chile, which appeared, unlike so many other countries in Latin America, to be on a path to join the world's wealthiest nations. Many other parts of the developing world, such as China, are also on track to lift their masses out of poverty. This rapid development makes today an exciting time to live. But for countries to achieve such wealth without dramatically altering the planet, we need to discover nonpolluting ways to improve standards of living. And that, to me, is one very clear way that we can do something to help the poor of the world rise out of poverty—make clean energy affordable, so the development that all countries want and deserve doesn't contribute to climate change.

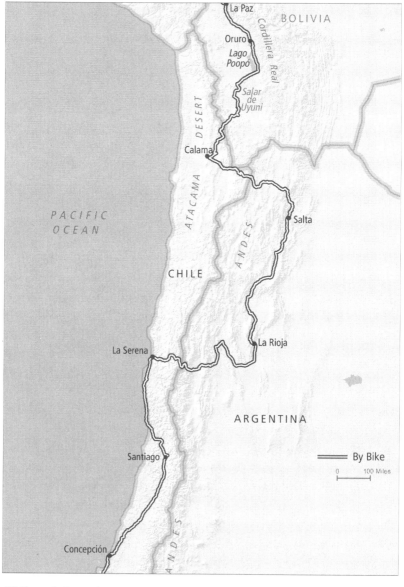

Chile and Argentina route

19 NORTHERN CHILE ڴ
The Miners

Chile population: 17 million
Annual per capita GDP: $22,400
Annual per capita CO$_2$ emissions from fossil fuels: 4.2 tons

DAY 392
December 1—The Atacama Desert

A fiercely cold and dry high-altitude wind blew at my face and then at my side. On a few stretches it was so fierce I pedaled in my lowest gear—riding on almost flat terrain. At times I was forced to dismount and push *del Fuego* forward along the sandy road. So little rain falls in this region that rivers haven't cut paths in the mountains; instead the water feebly collects at the bottoms of unconnected saline basins. Pink flamingos waded in some of these salt pools. *What were those birds doing there?* I thought. *What was I doing here?* I thought.

At 6 PM, pink strips of sunset reached across the volcano peaks. Though most of the red giants were dormant, thin smoke billowed up from the top of one cone—proof that the Nazca Plate was still diving under South America, pushing these mountains higher. Rounding a salt flat, surrounded by these peaks, I passed the first human structures I'd seen since a tiny town at the border between Bolivia and Chile. A series of mobile homes sat just at the edge of the salt flat; on the salt flat sat an idle bulldozer and a pair of wooden soccer goalposts with nothing but leveled salt between them. The goals looked out of place, like the golf

313

clubs astronauts took to the Moon. Longing for shelter from the cold wind, I rolled up to one of the mobile homes and knocked.

The door was answered by a man who introduced himself as Valerio. He had crooked teeth and a red baseball cap, and looked as dirty and lonely as I did. He said little more than to invite me to cook my dinner of pasta, onion, and egg in the small room. I asked Valerio what they did here.

High Atacama Desert

"We mine borax," he said. "Normally there are twenty other people here. Right now it's only me. Others will come in another few weeks." He worked off and on, three weeks at a time: on for three weeks at the mine, and then off for three back at home in Bolivia. "It is easy to find work in Chile," he said quietly.

I asked if he had a family. "Three children," he said. "I have a daughter at university in Sucre, studying accounting," he told me, and I thought of the mud-stained faces of children staring at me from the schools of Bolivia, wondering how many of them got to go to university. I asked if it was lonely here. "Yes, sometimes," he said. I shared some of my food, and we sat at a small circular folding table with Coca-Cola branded plastic chairs. Valerio then let me nap on his floor. After an hour, nourished by both food and brief companionship, I donned three layers of clothing and returned to the road to ride in the moonlight.

I sought my physical limit—I had biked late into the night the day before as well. I kept expecting to have an out-of-body experience, much like I've heard some cyclists experience when competing in masochistic events like The Furnace Creek 508 or Race Across America.

I imagined exhaustion would free me from my body, and I'd watch myself pedal from above. I'd see myself biking along the empty dirt road, weaving through the volcanoes that looked like a Martian landscape. As the night wore on, thin clouds eventually covered the Moon, and the mountain silhouettes became darker, more dreamlike. I could barely make out the dirt road ahead, and I biked almost by feel.

I encountered no out-of-body experience and achieved no enlightenment—I was just extraordinarily tired. At 3 AM I threw my sleeping bag in the dirt fifty feet from the road and collapsed.

In the morning I left the terrain of high-altitude salt flats and descended the western slope of the line of volcanoes. The infrequent sagebrush gave way to only dirt and rock, a landscape completely free of vegetation. The Atacama Desert is routinely referred to as the driest place on Earth, with some portions not seeing precipitation for many years. I've heard there are adults in this region who have never seen a real rainstorm. While most deserts will probably get drier with climate change, scientists aren't sure what will happen to this one, although it's hard to imagine this desert getting any drier.

Dropping down, the road eventually met a small stream that squeezed a trickle of water from the mountains. Vistas of sand and dirt were replaced by buildings on the outskirts of Calama. I passed a series of houses, boxy, modern structures painted white and yellow, each with a garage and surrounded by metal gates. The architecture surprised me; other than the borax mine, the last settlements I had passed through were mud villages where Bolivians grew quinoa and raised llamas in the desert.

Hungry and tired, I pedaled to a mall on the city's outskirts, navigating a large parking lot where cars shone in the desert sun.

DAY 393
December 2—Calama

I'd been to Calama before, six years earlier. I had studied in the capital of Santiago for three months during college, and once took a twenty-four-hour bus ride to the desert, spending a few hours in this city. The mall had been built since then.

"Can you watch my bike for a minute?" I asked a security guard standing by the automatic door.

The guard stared at me as if looking at an alien. This was different from stares I had received in remote villages where people don't often meet travelers—it was something else, as if he wasn't sure whether to trust me.

"I just want to go inside for a minute and buy some food," I said.

"Okay," he answered. "I'll watch it."

But still, why had he stared at me? I walked into the mall unprepared for what awaited.

Christmas lights hung from the ceiling, and a two-story-tall decorated pine tree stood in the middle of the mall. A store window displayed a line of plasma TVs. Next door another store sold new Chevrolet sedans, minivans, and SUVs beneath a poster reading, in English, AN AMERICAN REVOLUTION. The abundance of goods surprised me. Chileans wearing jeans, sneakers, and T-shirts walked the mall, carrying their purchases in plastic bags.

I blinked again, confused. The last "mall" I had visited was an outdoor market in Bolivia that sold live chickens. In one way, it felt strangely comforting to be here, as if I were back in the States. But that same sensation also felt deeply disorienting—I was supposed to be on the far side of the Earth.

Then a couple stopped and briefly stared at me much the same way the security guard had. It eventually dawned on me that my appearance was as out of place to them as the mall's First World materialism was to me. I hadn't shaved or showered in a week, and my face was covered in desert dust. I suddenly felt as if I was offending everyone in Chile. I purchased a Coke and candy popcorn from a small booth and quickly made for the exit.

Despite my appearance, the Calama firefighters welcomed me for the night. Fit for company after a shower, shave, and nap, I sat and chatted with the Chilean *bomberos*. My take on this station was similar to that of the mall. The last fire station I'd stayed at, in Oruro, Bolivia, used a truck from the 1960s that belonged in a museum, whereas Calama's truck was new, and the three-story station was freshly painted blue. The city's tap water was even safe to drink.

But still, these *bomberos* were volunteers—all worked at the nearby copper mine. Chuquicamata is the world's largest copper mine; miners removed an entire mountain to create its open pit. A *bombero* named Pedro walked me over to the station's window and pointed with pride to the brownish yellow mountains on the horizion—no vegetation in sight.

Shopping in Calama

"You see the horizon over there, where that mountain is flat," he said to me. "There, between those two mountains, is a pit almost a thousand meters deep. This whole town is built on copper wealth."

"And the price of copper is way up, right?"

"Yeah, it's increased over 200 percent in the past five years." Pedro opened a laptop and showed me pictures of a giant pit two miles across, its edges built like a giant staircase. The staircase was actually a ramp circling the mine, by which dump trucks carried out the raw material to be processed.

Pedro confirmed that the mall I'd visited was newly built, as were a hospital and many of the housing developments. The local economy was definitely doing well. He had moved there from Copiapó, a city to the south, to take advantage of that opportunity. "The desert isn't the best place to live, but, well, the pay is good."

Pedro then excused himself, saying he needed to study. He opened his laptop on the nearby table. Two other *bomberos* did the same.

"Study for what?" I asked.

"There's new mining equipment—we have to pass a few tests to use it."

"Do those laptops belong to the mining company?"

"No, they are ours."

———————

Over fifty years before, Ernesto "Che" Guevara, crossing Latin America by motorcycle, had also met mine workers in Calama—but he had a very different experience from mine.

Che's journey, undertaken well before he became a leader in the Cuban Revolution, is chronicled in his book *The Motorcycle Diaries*. In the book Che told a compelling story of the poverty experienced by mine workers near Calama. He met mine workers who had just been released from prison for having partaken in a strike, and who were struggling to make money and survive. He described them as "a live representation of the proletariat of any part of the world." Some scholars consider this a pivotal moment in Che's journey toward socialism.

During my visit, on the other hand, the mine workers sat in the fire station browsing the Internet on their laptops, reading over PowerPoint presentations, comfortably huddled around the glow of their computer screens. Perhaps these firefighters are the exception, and many of the mine workers are likely still poorly paid. But, while poverty still stretches across much of Latin America, this region at least is far wealthier and better off than when Che crossed it.

By almost every measure, the quality of life of the average Latin American—someone living between the northern border of Mexico and Tierra del Fuego—has dramatically increased in the past half century. The average income in Latin America is almost three times higher today than it was fifty years ago, the infant mortality rate has dropped by a factor of five, and the average Latin American born today can expect to live almost twenty years longer than his or her counterpart born in 1950.

The title of my book echoes Che's journey; *The Motorcycle Diaries* resonated strongly with me. He poetically described what it was like to live off the generosity of strangers, to ultimately empathize with an entire continent—I certainly feel the same. And, like him, the "goal" for my journey wasn't always clearly linked to what I was doing. Che was ostensibly looking for a cure to leprosy; I was raising awareness of climate change. Both of us appeared in the media, and then received assistance from others because of that publicity—I can see that sometimes, to an outside observer, it wasn't clear if we sought the media in order to promote our mission or just to further our travels.

But there is a lot our respective stories don't have in common. When Che witnessed the oppression of the underclasses, he deemed it necessary to remake the world's economic system—violently, if need be—in order to lift them out of poverty. But I don't reach the same conclusion. In the South America I crossed, there is indeed great poverty and inequity, but people's lives in many regions are improving without the type of revolutions envisioned by Che. Yes, many communities are still trapped in poverty. During my journey I saw homes that lacked even outhouses, and families that survived at the mercy of the rains. I biked through slums of mega-cities where the quality of life actually frightened me. Such inequity is still offensive. But I also encountered televisions and Internet in the poorest of villages, a kid's portable DVD player in the rural forest of El Salvador. I met a Bolivian man whose employment mining borax in Chile enabled him to send his daughter to university. The *bombero*-miners of Chile owned laptops—and had workweeks that allowed them free time to volunteer. These examples and more left me thinking things are on an upward trend. I am hopeful that, in the next fifty years, though perhaps not without difficulty, Latin America may yet be able to eradicate material poverty.

NORTHERN CHILE
December 1–6
Presentations: 0
Flat tires: 0
Miles: 343
Trip odometer: 11,925 miles

20 NORTHERN ARGENTINA ⮷
Finding a Route

Argentina population: 41 million
Annual per capita GDP: $11,500
Annual per capita CO_2 emissions from fossil fuels: 4.5 tons

DAY 398
December 7—Across the Andes

My chosen highway would lead me into the Andes again—an 8,000-foot climb followed by a hundred miles of desert with no stores for supplies, no streams for water. So I strapped ten liters of water and two days' worth of food to the bike, and spun uphill in my lowest gear beneath a full Moon, cycling into the freezing desert night carrying the heaviest load of the trip. I could have followed a more direct route, continuing south through Chile, but northern Chile is stark empty desert, and I preferred to cycle the cities and forests of northern Argentina. The Andes divide the climates of Chile and Argentina in an interesting pattern. Andean topography is such that it blocks precipitation in the northern part of Argentina from traveling west to Chile—so northern Argentina is lush while northern Chile is barren. But in the far south of the continent, more than 1,500 miles and 20 degrees of latitude from where I was—consider the range of temperatures between Seattle and Mexico City—the wind directions and thus the climates are reversed; the eastern Argentinian side is arid steppe, while the western Chilean side is temperate rainforest.

But the climate along the Andes can also change dramatically within relatively short distances. As a result, the best conditions for riding were not to be found with a single route. Sometimes better riding was on the eastern side of the mountains, sometimes on the western. To accommodate, I planned to cross the three-thousand-mile Argentine-Chilean border perhaps as many as seven times before reaching the tip of the continent. Each time, I switched the flag on my rear rack from Argentinean to Chilean, and then back again.

Del Fuego by moonlight

I mentioned earlier that I had planned my entire journey following the calendar, timing my ride so as to avoid the rainy seasons as well as the extremes of winter and summer. So I largely avoided the three great belts of rain that circle the Earth: one near the equator and two others located somewhat symmetrically in the temperate regions to the north and south. In between these is drier weather, especially in the subtropics, which is why most of the world's deserts are found between 20 and 30 degrees latitude.

As the seasons change, the direct light of the sun moves from 22.5 degrees north of the equator to 22.5 degrees south of the equator, during which time the belts of rain also vacillate north and south. This movement is why California is bone dry in the summer but wet in the winter—the rain from the far north in Canada and Alaska moves southward during the winter, and then back northward during the summer. Likewise, the rain belt at the equator moves southward during the winter, drying out Mexico, but then northward during the summer, when Mexico and Central America receive their precipitation.

So I plotted my trip such that I'd left California in November—just escaping the beginning of California's rainy season—and then biked Mexico and Central America during their dry winter months, when the equatorial rain was farther south in South America. This rain belt moved north as I biked, and I passed through it in Colombia and Venezuela. Now, as the southern hemisphere summer approached, that same belt of rain was moving south—and I was riding just in front of it, avoiding rains in both Peru and Bolivia. If I had passed through Bolivia just a month later, the salt flat would have been covered with a thin layer of water.

If you were to envision these belts of rain moving with the seasons, and then look at my route, you would almost see the dance I enacted—continually working to avoid the rainy seasons and to bike into the most optimal weather. Now, if I were to do this same journey again in fifty or a hundred years, would these large-scale rain patterns have changed? Would I have to plan my route differently as a result of climate change?

Such a route will not necessarily change much. There will still be summer and winter, and the rainy seasons and dry seasons will continue to oscillate. But the rain belts may shift. The belt found near the equator, the Intertropical Convergence Zone, or ITCZ, does not follow a straight ring around the globe—the shapes of the continents cause it to snake somewhat north and south. Some researchers project that within this century the ITCZ might not travel as far north during the summer months as it does now; if this happened, Mexico and Central America wouldn't receive their normal rains. Such a weather pattern probably persisted during the drought that coincided with the fall of the Maya Civilization.

Global warming adds heat to the engine that drives these weather patterns. And with more heat comes more moisture in the atmosphere. Nearly every climate model suggests that rainfall will thus increase both along the ITCZ near the equator (though the ITCZ itself may move) and also along the higher latitude rain belts. On the other hand, rainfall will likely decrease in the subtropics—the dry region of deserts just to the north and south of the ITCZ, such as the American southwest, much of Mexico, and central Chile.

Although regional trends are difficult to measure, we can already detect these changes. The subtropics are already slightly drier, and the higher latitudes are receiving, on average, more rain than they did fifty or a hundred years earlier. Though the trends so far are barely discernible statistically, they seem to be following the climate models' predictions.

DAY 398
December 7—El Payaso

One day after crossing into Argentina, where I got my passport stamped at a lonely customs building, I took a brief break before noon, resting *del Fuego* on its kickstand and eating bread and cheese beneath the deep blue Altiplano sky. At one point I looked back, and saw three cyclists approaching me on the empty road.

I laughed and yelled: "Hurry up! I'm kicking your butt!" It was Gregg and Brooks—whom I hadn't seen since the jungle of Guatemala, 40 degrees of latitude and ten months earlier—plus a third rider who'd joined them. There I was, in the middle of a vast desert at the top of a remote part of the Andes, and yet friends could still cross my path.

"Look," said Brooks, making a wry smile, "It's El Payaso." It seems I would never be able to live down that day of errant sunblock and salsa.

They pulled over and dismounted. Their sponsors had long ago replaced the gear stolen in Chiapas, so they carried four black Ortlieb panniers just as before. Though I'd been in sporadic contact with Gregg via email, I thought he and whomever he was riding with were many days ahead of me; I must have recently leapfrogged them, and now they'd caught up with me. Gregg and Brooks had been joined by a third cyclist, Tom from Colorado, who pedaled a mountain bike with a long single-wheeled trailer much like Wouter had. Tom had started in Ecuador, and was also on his way to Argentina.

Although we had met by coincidence, I somehow wasn't surprised: we were following more or less the same route, so I'd assumed we'd cross paths again at some point. Or, at least I can say that about Gregg. I was surprised, though, to see Brooks, who had flown home after Guatemala and returned to his old job for six months. He had then flown back to southern Peru to bike with Gregg for two months before returning to get married. In the meantime, Gregg had biked the entire way.

So we set off together in the thin air of the Altiplano, riding side by side and chatting on the traffic-free road. Though we were slowly approaching the eastern, wetter side of the mountains, the terrain was still dry, largely empty of life other than sagebrush-like plants. This region, the northern border between Chile and Argentina, was remote; from there to central Chile, a distance of a thousand miles, only four roads crossed the Andes—only two of which were paved, and none of which had significant traffic.

Gregg described their most recent adventures. They had taken a remote dirt road through southern Bolivia, then spent a few days pushing their bikes uphill on sandy roads. As Gregg had also biked Peru, we swapped stories of our favorite dirt roads in the country. Every long-distance cyclist has a different style of riding, and thus a slightly different experience. I had stayed at fire stations and promoted an environmental project; they had stayed largely in hotels and focused on their photography. Yet we still saw the same thousands and thousands of miles of countryside.

Down the road a while, Brooks and Tom a few hundred yards behind, Gregg and I talked about what happened after we'd parted in Guatemala. Brooks had written a detailed and well-thought-out email to their supporters about why he was going to avoid Central America, especially how it wasn't safe. But then Gregg wrote the same supporters to say he was continuing. "All of my friends, convinced by Brooks's decision, told me I was crazy and reckless. I lost my moral support. And then the poverty of Central America really got to me. Central America was tough."

"How do you think you made it through?"

"Once I made it to South America, I felt like something had changed. But also, you know, when you set a goal that is so big, so enormous—like biking to Tierra del Fuego—it pulls you toward it. Every day I know where I am headed, because that huge goal is pulling me forward. It's like what happened to you in Mexico. I tell that story all the time."

"What do you mean?"

"Remember how you didn't know anyone in Mexico, and then you emailed that one guy in Durango? He got you in contact with that guy in San Luis Potosí who then got you in contact with the bicycle group in Mexico City, which then got you on the national news and got you contacts all the way south. You've been riding that wave the entire journey."

I thought back to Baja California, when I knew absolutely no one. A single email had led to a series of contacts that grew and grew. Gregg continued: "For me, once I decided I was going to ride to Tierra del Fuego, I suddenly found all these other cyclists who had done the trip and who gave me advice, and then Co-Motion bikes got behind us, and one thing led to another, and it's felt like the goal has been pulling me instead of me riding toward it."

I asked Gregg what he would do when he finished the trip—what would be his "Tierra del Fuego" after reaching Tierra del Fuego.

"I don't know," he said, "but I know I can't go back to my old job of selling software. That life is over." After a pause he asked, "How about you?"

I said I didn't know. I hadn't thought through what I'd do after I finished riding. That is, after I biked across the U.S. after reaching Tierra del Fuego. It seemed so far in the future. I'd assumed that, if I gave this ride and project all my effort, the next steps in life would be obvious. We rode in silence for a minute. Then I changed the topic, asking a question that had been on my mind, "What do you think of the poverty that you've seen? Had you seen poverty before this journey?"

"I'd seen it, but it was always through a bus window. It's so different on a bicycle, when you're in a town and trying to fill your water bottle. The people living there are real. It makes me sick to my stomach to see the poverty."

"I also feel sick," I replied, "but I'm confused as well. I've also seen people who seem to have happy lives and supportive families." I thought of the shrimp farmers in southern Mexico, or the family of Seventh-day Adventists in Peru and their early-morning prayer. "The difference is that they lack options. And that's what makes me feel sad. They can't do anything near what I can with my U.S. passport and a first-rate education."

"And you know what?" said Gregg. "They know that. I was talking recently to a woman in Peru. She said, 'You gringos, you think that we don't know about what you have, but we know much more than you think we do.' I mean, even in the smallest villages they have Internet now. They see our movies. They know there is a world of wealth out there that they can't have. I think it results in frustration. Wouldn't you be frustrated if you knew your options were limited? If what you saw on television were impossible for you to obtain?"

We rode in silence for a bit longer, and then reverted to complaining about our lack of success with women. The paved road led us up over a final 14,000-foot pass, and then we descended through a canyon, arriving for the first time in over a month in a landscape of forests and rivers. Thick clouds hung overhead, and trees with broad leaves lined the road. I hadn't seen a terrain this green since the jungle of Peru more than three months earlier.

Jujuy, a city of perhaps a few hundred thousand, had small cars and busy traffic and narrow streets, part of a city grid laid at least a hundred years before. Few of the cars or buildings were new. The old wealth—stately colonial buildings—contrasted sharply with the poverty of Bolivia and the new shopping centers of Chile.

Argentina was once considered a wealthy nation. At the turn of the twentieth century, the average Argentinean was as wealthy as the average person in Western Europe or the United States, a prosperity that can still be seen in the fin-de-siècle architecture of the major cities. Yet today a citizen of Argentina is less than a third as well off as the citizens of the United States. Though people speak of Argentina's "decline" over the past century, it's more accurate to state the country's economy merely grew more slowly than that of other wealthy countries. The average Argentinean today is still much wealthier than the average Latin American—as well as much wealthier than the average Argentinean was a century earlier.

About a mile into Jujuy we stopped at a corner store, where we bought fresh bread, salami, cheese, and cheap wine. At a nearby café sat elderly men with fair complexions, smoking and sharing stories. At noon, the stores closed for a three-hour siesta. "Are we in Europe?" Brooks asked.

DAY 403
December 12—Christmas in the Desert

In the city of Salta we parted ways again. Gregg and Tom would take a few weeks off in town, and Brooks would return to the U.S. We said goodbye in the way we always did—assuming we'd run into each other again, but neither planning to nor counting on it, as if we had faith that the journey would determine what was best.

As I rode south from Salta, the terrain become browner as I cut along the foothills of the Andes. I continued on with a strange mix of emotions, simultaneously being weary of travel, wanting to proceed as quickly as possible, and yet also not actually wanting the journey to end.

It so happened that exactly the friend I'd want to join me for this section was flying into Santiago, Chile, and then taking a twenty-four-hour bus ride in order to ride with me: Dave J. Dave and I, like many guy friends, became friends through shared athletics—and he was also the type of person who liked to push himself to see how far he could go. The result was that my daily mileage jumped once he joined me; where I had averaged 50 to 70 miles each day previously, Dave and I covered more like 80 to 100 miles, drafting off one another. As I was for the time being freed from any Ride for Climate agenda—schools were on vacation and large population centers were few and far between—our only goal was to put on miles.

Dave J. biking into a sand storm

With daily temperatures in the 90s, we rode southward through the desert for a week. The desert offered little shade or wind protection; at one point a sand storm blasted grains into our gears and onto our skin. We spent Christmas on an empty dirt road, which we followed for three more days before seeing a store. This road led us back over the Andes to Chile, crossing a 15,000-foot pass through brownish red mountains with small glaciers. Dave, not being as well acclimatized as I was, cursed the altitude and dirt roads, but his companionship nonetheless helped me get through another long stretch of road. Dave then returned to the U.S. Having sent off my last biking companion in Latin America, I again headed south, continuing my journey alone.

Crossing the Andes, again

NORTHERN ARGENTINA
December 7–25
Presentations: 0
Flat tires: 5
Miles: 1,029
Trip odometer: 12,954 miles

21 CENTRAL CHILE ☼

Prosperity & Pollution

DAY 422

December 31—New Year's Eve

Is this California? I thought. I felt as if I'd been teleported home—or maybe I had biked a full loop around the Earth, arriving back in the Golden State.

Chile's coast

Tufts of grass clung to the tops of rocky cliffs, and a haze hung over the horizon, blurring the meeting point of sky and sea. A cool breeze lofted salty fog, and along the roadside grew purple thistle just like I'd seen in California. *Avena barbata*, a grass with an oat-like head, and erodium, a small annual plant with a Dr. Seuss–like pointed seed, grew next to the roadside; I recognized both plants from Jasper Ridge Biological Preserve in California. Even the roads reminded me of home. Chile's rapid economic development had brought freshly paved highways, along with the occasional SUV.

Central Chile, which spans roughly the same distance north to south as California, is also about as far south of the equator as California is north. And like California, it has a rocky coast along the Pacific and tall inland mountains. Central Chile's weather is also similar to California's: a rainy winter and a bone-dry summer. In the winter, snow accumulates in the Andes; come spring and summer, the melting snow releases water for the country's copious orchards and vineyards. As I had arrived during the southern hemisphere summer, I encountered many roadside stands selling peaches, avocados, and grapes.

Unsurprisingly, global warming will challenge central Chile in many of the same ways it will challenge California. Because the summer is dry and the Andes accumulate thick snowpack in the winter, cities and agriculture rely on snowmelt for their water supply. But climate models suggest that both central Chile and California will likely see decreased precipitation in a warmer world. Each country's "mirror image" on the map will have similar problems: drought and agricultural losses.

On New Year's Eve, about two days north of Santiago, I rolled *del Fuego* onto a sandy beach between low sea cliffs, nestled beneath a haze created by cool ocean currents. It appeared to be a sort of informal beach campground: a few large tents stood on the beach, a few SUVs parked nearby. The tents were of varying quality. One, mostly in tatters, belonged to a fisherman listening to a crackly radio. I chatted with him briefly, then walked over to another cluster of far newer tents set up near a boxy silver van.

I was waved over by a woman in her fifties, who introduced herself as Petra. After I explained my journey, she exclaimed, "Come join us for a New Year's celebration!" I then met her husband, Pedro, three of her children, and three of her grandchildren—all of whom had curly black hair and the relaxed smiles of people on vacation. Their camp was comfortable and elaborate, with a series of tarps forming small enclosures. One enclosure had a mini kitchen, complete with portable gas stove and hanging pouch of tomatoes, avocados, and peaches. Under another tarp was a plastic table draped with a fresh cloth. Nearby a four-foot-wide solar panel charged a large battery.

After setting up my tent about twenty yards away I chatted with Pedro, who told me they were spending nine weeks on the beach. Since it doesn't rain during the summer, they could live there quite comfortably. "I just retired last year after thirty-five years serving in the army," he calmly told me.

"Wow," I said, quickly doing the math, then blurting out: "You were in the army in 1973." That year, the army general Augusto Pinochet had violently seized power, overthrowing a democratically elected government and installing himself as dictator. He then ruled the country for nearly two decades.

"Yes," Pedro replied, without changing his tone or showing unease.

New Year's Day breakfast

"What was that like?" I couldn't help but ask, somehow not realizing it might be inappropriate to inquire about this period of Chile's past.

"There were injustices," he said flatly. "There were many injustices that I don't approve of—things that I did not learn about until recently.

But in all, it was worth it." In hearing that statement, I couldn't help but recall how, in the process of seizing control of the country, Pinochet had imprisoned perhaps eighty thousand Chileans, tortured as many as thirty thousand—even women and children—and killed two to three thousand.

"What do you mean?" I replied.

"Pinochet saved us. If not for him, we would've been another Cuba. Instead we have a system that works. Thanks to Pinochet."

"You know," Petra said, overhearing the conversation, "you always hear about the injustices of the Pinochet regime. But the other side also fought. Many army officials had their houses shot at. My friends feared for their lives."

Although I know a somewhat significant portion of the Chilean population holds the same opinion as Pedro, I had a difficult time believing that the injustices of the opposition were as bad as Pinochet's. But as a guest, and as someone who didn't live through Pinochet's rule, I stayed quiet. I was also thankful to have people to spend New Year's Eve with.

Pedro cooked large strips of beef cooked over a grill for dinner. Their solar-powered battery provided light once the sun set. His teenage son sat by himself and played with an MP3 player; later he took a picture of me with a digital camera. I was impressed by the electronics, as well as by the quality of the camp and the abundance of high-quality food. We shared a large meal of potatoes, tomato salad, and grilled beef, and Petra saw to it that my glass of wine remained full. The family's affluence—and the quality of the wine—made me again ask myself if I had returned to California.

Economists sometimes speak of the "miracle of Chile" when talking about the country's remarkable economic growth. From 1975 to 2005, Chile's per capita income grew by an average of nearly 4 percent per year, such that today average Chileans are three times wealthier than they were in the mid-seventies. This economic growth is far and away the fastest in Latin America; no other Latin American economy has grown half as fast in the same time period. In fact, in some countries—such as Venezuela, Bolivia, Nicaragua, and Peru—the average citizen was *poorer* during my journey than they'd have been thirty years earlier.

Some, such as Pedro and Petra, credit Pinochet with this economic progress. After overthrowing the democratically elected socialist government in 1973, he completely reworked the economic system. Whereas the previous government had nationalized or tried to nationalize countless companies, Pinochet privatized them; thereafter Chile rapidly transformed from an economy run by those on the far left to those on the far right. Of course, it's difficult to credit Pinochet entirely for Chile's growth: the economy collapsed in his first decade of control, and the period with the most rapid economic growth occurred after he handed over power to a democratically elected government, which was run by a center-left coalition from 1990 until 2010. Whatever the cause, though, the long-term trend of the economy has been strongly and undeniably upward.

This economic growth has not come without challenges. Much of the growth has been fueled through the depletion of natural resources, such as copper in the desert north and timber in the rainforest south. The far south of Chile is temperate rainforest, much like the Pacific Northwest of North America; huge tracts of native forest have been cleared to make way for plantation farming. Income inequality has also increased during the past three decades. While it's true a majority of Chileans have benefited from economic growth, the wealthiest Chileans have benefited the most, whereas many others, like the fisherman on the beach next to Pedro's tent, have not seen their incomes grow as substantially. Nonetheless, for most objective measures of quality of life—including income, education, and health—Chile has improved dramatically. Indeed, some Chileans are like Pedro—retiring early, buying expensive solar panels, and spending time on the beach with their grandchildren.

DAY 423
January 1, 2007—Santiago

The following day the highway led me inland through brown grass-covered hills. Farther on I passed a series of orchards irrigated by water from the mountains. After riding sixty miles, I camped next to a house whose occupant offered me ten avocados. "We have too many," she said, "and they will go bad." I looked at the fruits and thought of how many times during the California winter I'd seen avocados in the supermarket labeled Product of Chile. That made me think of how I often disagree with those who argue for eating locally for environmental reasons. For one, shipping makes up only a small part of the environmental footprint of our food; and two, it's meaningful to me to economically support people living at the far corners of the globe. Perhaps more strangely, when I now see Chilean fruit at the supermarket I actually think of it as local. After all, the orchards are just a bike ride away!

I continued south on the freeway, the only route into Santiago. Traffic increased as I approached the capital, but a wide shoulder offered a safe buffer from the cars, even in a few spots where signs stated bicycles were prohibited. I rode on, through dry hills and finally into Santiago, where a third of the nation's seventeen million people live. Santiago sits in a valley that's a condensed version of California's Central Valley, with the peaks of the Andes rising up to the east and the more gentle, oak-covered hills of the coastal range to the west. I felt like I was riding home.

I had no map of the city, but I didn't need it; I'd lived in Santiago for three months during college, so I had no trouble finding my way. My destination: the house of my old host family in Las Condes, an upper-middle-class neighborhood on the far side of the city.

Following a bike commuter in Santiago's Providencia district

Following the route of the Vespucio Norte metro system, I biked through low-traffic neighborhoods of single-story homes. After passing the zoo I arrived at the Río Maipo, a river colored brown from glacial runoff and held in place by concrete walls. A freeway—a new freeway?—lined the north side of the river. Much as I'd been surprised by the progress of Calama, I was surprised by new roads and buildings in Santiago. Many new structures had sprung up in the six years since I'd been there.

I crossed a bridge over to the Parque Forestal, where broadleaf trees arched over the park's grass and displayed the thick green of summer. Beyond the park stood the tall buildings of Providencia, one of Santiago's several municipal districts. The tallest in the city, the Edificio Telefónica, was built to appear like a giant cell phone, with a smooth glass front and a tower at its top like an antenna.

Dodging trucks, buses, and taxis, I followed the main road through Providencia, passing tall buildings and cranes constructing more tall buildings. Then an Ideal-brand bread truck spewing diesel fumes suddenly turned in front of me, almost running me over—I avoided disaster thanks to my well-honed reflexes. But even so, compared to other capitals I'd biked in, this was not chaotic traffic. The four lanes of newer cars appeared more organized, as if this were a country where people were more likely to follow the rules.

In the residential neighborhood of Las Condes, each small lot was divided by black metal fences and well-tended gardens. Amongst summer blooms the lawns were meticulously trimmed, the bushes pruned with care. A few dogs barked at me from their fences. I walked *del Fuego* along the street and stopped in front of the Bravos' house. The two-story home looked unchanged, but a new van was parked in the driveway. The door cracked open in response to my knock.

"Look, it's the gringo *ridículo,*" said Miriam, my host mom, looking smugly at me without surprise and placing her hands on her hips. The "ridiculous gringo"—that was what they called me. Or Señor Energía: Mr. Energy. I suppose I had remained true to those names by returning on bicycle. "Come in," she said, with a demanding smile.

She offered me *once,* a Chilean meal of bread with jam, avocado, or cheese served with tea or coffee, enjoyed at around 5 PM, a few hours before dinner. As we'd been in only sporadic contact since I'd lived there, Miriam filled me in on the past half decade, explaining how her daughter and son had finished school and were working, and how Juan, her husband, still operated a taxi service, but now with a new van. Then she asked what I'd been doing with myself.

I explained how I'd earned a master's in environmental science, worked in a laboratory, and embarked on an educational bicycle adventure. "But what will you do when you are done biking?" she asked. I told her I'd figure something out.

I offered Miriam the ten avocados I'd been given the day before, telling her I'd just carried them a hundred miles. She laughed and thanked me. "What a ridiculous gringo," she said, and showed me to my old room in the small two-story house.

When I was first in Chile, six years earlier, I had been impressed by the scale of globalization I found there. Chile had KFC, Domino's Pizza, McDonald's, and even smaller chains like T.G.I. Friday's. It was then the dawn of the new millennium, and everyone was talking about globalization—was it good or bad? What type of globalization did we want? Did globalization mean Americanization? I had also been impressed by just how closely Chileans followed U.S. politics—as it was the year 2000, every Chilean I met had an opinion over whether Gore or Bush should win for U.S. president. (I also understood why they would follow United States politics; the CIA significantly aided the Pinochet coup in 1973, and all across Latin America—and the world—the U.S. government has continued to interfere with regional politics.)

During this second visit to Chile, though, my biggest impression was not the country's relations with the United States or the state of globalization, but rather its general scale of progress. I didn't expect high-definition televisions hanging over the subway platforms. I didn't expect all the new cars, the new asphalt, the new highways, the new buildings. In Providencia I'd seen cranes in the process of constructing La Gran Torre Costanera, which when completed will be the tallest building in Latin America. (Chile, like California, has violent earthquakes, making the building's construction just that much more audacious.) In the mere six years since I'd left, the country appeared to have progressed twenty years.

With this growth, though, came increased pollution. Since 1990, the country's annual fossil fuel pollution has increased by about 50 percent, from about 2.6 tons to nearly 4 tons. Despite this growth, Chile's per capita carbon dioxide emissions aren't particularly high by Latin American standards, and are only one-fifth of what the average U.S. citizen produces. The reason is partly because Chile derives a large portion of its electricity from hydroelectric dams. Nonetheless, its per capita electricity use has more than tripled since 1980. So, though Chile's increase in pollution isn't exorbitant, it's still worrisome.

But on the other hand, it's exciting to think the poor of the world could achieve what Chile is accomplishing—that the many slums I'd biked through could soon afford better infrastructure, that electricity could someday reach every home. Fortunately, much of the developing world is actually developing. Though a handful of countries—mostly in sub-Saharan Africa—remain stagnant, the majority of the world's population is experiencing economic growth.

But, inevitably, as economies grow, so will pollution. And in the next few decades, almost all the growth in carbon dioxide pollution will come from developing nations such as the ones I biked through. The International Energy Agency estimates that the growth in developing countries will increase the global demand for energy by about 50 percent in the next twenty-five years.

Chile's story does offer some hope, though, as the country is already pledging to reduce carbon dioxide pollution. Soon after my trip ended the Chilean congress passed a bill pledging that by 2014 Chile would produce at least 5 percent of its energy from non-hydroelectric renewable energy, increasing that amount to at least 10 percent by 2024. (Note that the president at the time, Michelle Bachelet, had wanted an even higher target, but she had to compromise with congress.) These numbers are significant. Clean energy is, after all, currently more expensive than traditional energy. The good news is that, once a country like Chile becomes wealthy enough to afford more sustainable resources, its citizens are more likely to invest in clean energy. But to get to this point they must first become wealthy. I always remember what the former president of Guatemala told me: developing countries "must focus on development," and they will use the cheapest energy to do so.

The problem, though, is that greenhouse gases accumulate in the atmosphere. Yearly pollution doesn't matter so much; it's the total cumulative emissions that cause concern. So let's say the entire world grows its economies with fossil fuels, and then switches to alternative energies once they're affordable—even if everyone eventually makes that switch, it might be too late. We will have already committed the world to too much climate change. It would not be enough to merely increase more slowly, or to stabilize the current levels. Our total emissions must *decrease* if we are to prevent disaster.

Early evening in Santiago

We may have only one option: making clean energy cheaper than coal and oil so all developed and developing countries adopt clean energy sooner rather than later. In some ways, what we have before us is a race. Which will develop first: the developing world, or clean energy that's cheaper than traditional energy?

Since the schools that were ideal for talks were on vacation, I instead spent time with my host brother and sister and their friends, enjoying Santiago with my Chilean peers. Then, after ten days and one television interview, I departed the city, following a wide shoulder on the same freeway south. Riding during the Chilean summer brought another joy: for the first time in the journey I had more than twelve hours of daylight. In the tropics, regardless of time of year, the sun almost always rose at 6 AM and set at 6 PM. But in the summer of the Southern Hemisphere, I had light until 9 PM. This suited me well; I took long midday siestas, and then biked until the daylight faded.

As I biked ever southward, the roadside gave way to tall evergreen trees and cone-shaped snowy volcanoes—much as if I were riding north from California into Oregon and Washington. The tip of the continent, the far end of the Earth, was but two months away.

CENTRAL CHILE
December 26–February 3
Presentations: 0
Flat tires: 0
Miles: 1,196
Trip odometer: 14,150 miles

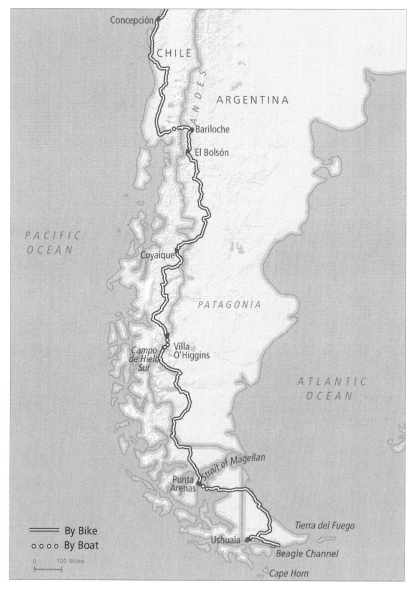

CHILE

ARGENTINA

A N D E S

Concepción

Bariloche

El Bolsón

PACIFIC
OCEAN

Coyaique

PATAGONIA

Campo
de Hielo
Sur

Villa
O'Higgins

ATLANTIC
OCEAN

Strait of Magellan

Punta
Arenas

Tierra del Fuego

Ushuaia

Beagle Channel

Cape Horn

―――― By Bike
ooooo By Boat

0   100 Miles

Patagonia route

22 NORTHERN PATAGONIA 武

Far from Home

DAY 457

February 4, 2007—San Carlos de Bariloche

I received the bad news after crossing into Argentina. I was in the tourist town San Carlos de Bariloche, which sits on a hill above the cold waters of Nahuel Huapi, a long skinny lake that reaches deep into the Andes like a freshwater fjord. I had spent three days without Internet, traveling dirt roads over mountains and taking a series of boats across lakes before arriving in Argentina. At a small café, I received an email from my mother.

> As you know, Grandma slid out of her wheelchair and broke her hip the first day I was here, a week ago. I have been with her in the hospital every day since. She was recovering well when suddenly she became critical: rapid heart, diving blood pressure. They were able to use drugs to get her heart working properly again, but she isn't bouncing back.
>
> Anyway, I wanted you to be up to date. Where are you now?
>
> Love,
> Mom

Grandma broke her hip a week ago? I hadn't even known she was in the hospital. She was ninety-one years old, the last of my grandparents still living, and the grandparent with whom I had been the closest. I had planned to fly first to Florida when I returned to the States so she'd be the first person I visited after reaching the tip of South America.

In the phone booth in the Internet café I made three calls: to my grandmother, to my parents, and to my older sister. No one answered. Anxiety worked its way deep into my stomach, and I felt my hands swelling. I took a deep breath, trying to tell myself she might be okay.

Leaving the café with *del Fuego* at my side, I walked across the cobblestone plaza. Bariloche is one of Argentina's most popular tourist destinations, especially during the winter, when a nearby ski resort draws international crowds. The majority of the tourists now were Argentine—most from Buenos Aires, on vacation, enjoying their summer. Young and elderly paced the plaza, smiling as if to taunt the cold Patagonian breeze. Rugged mountains, whose tops had been rounded by past ice ages, surrounded the town.

I decided that worrying about my grandmother wasn't going to help anything, so, though distracted, I conducted my outreach efforts regardless, falling into the routine repeated in countless cities. I returned to the Internet café, where I spent an hour writing and uploading photos. I got directions to the local newspaper, where I was interviewed by a reporter; I was then sent to the radio station next door, where I was interviewed for Radio Patagonia. I was no longer surprised by the ease with which I secured media attention—I just accepted it, my mind and heart only half engaged in the process.

As dark enveloped the town, I returned to the Internet café and called my parents. They didn't answer, so I called my sister. My brother-in-law answered.

"David, I'm sorry, I hate to be the one to tell you this . . . I just don't know what to say, I'm just sorry . . ." My heart sank. "She passed away this morning."

I walked *del Fuego* to the plaza, now enveloped in darkness. The cold wind beat against my Bomberos de Medellín jacket. I pulled a hat of thick Chilean wool over my head and sat on a bench, alone with *del Fuego*. I was in shock. I had known that Grandma didn't have much more time left, as her health had deteriorated over the past five years, but I'd thought I'd get to see her once more. I thought of everything my grandparents had done for my sisters and me—from Christmas gifts to helping with college—all given with unquestioning support. I sat for half an hour, staring at the silhouette of the lake and surrounding mountains, the wind drying tears from my face.

Needing a place to sleep, I asked for directions to the fire station, then rode to the far side of town. But I was told I couldn't stay because I wasn't "a real *bombero*"—despite my two uniforms and collection of *bombero* patches. Defeated, I rode to a youth hostel, where for ten dollars I got a bed in a shared room with Israeli and German tourists whom I had no interest in talking to. I felt even more alone than usual.

Lying down, trying to sleep, I wanted nothing more than to be home, holding my mother, experiencing our loss together. Why couldn't my grandmother have lived just two more months? Why hadn't I known she was in the hospital this past week? I could have called—I could have heard her voice again. Thinking of how I hadn't talked to her in months, intense guilt swept over me.

In the morning I called my parents, battling tears as I spoke. My mother said the service would be held in two weeks. "You can't wait till I return?" I stammered.

"Well, dear, you won't be home for two months."

The decision seemed to make itself. There was no practical or economic way for me to both attend the service and finish the project, and my parents didn't even request that I come home. "I'm so sorry," my mother said. I hung up the phone, feeling horrible, empty.

My grandmother was born during the Great War, and lived through the Depression and the Second World War. She was an impressive woman—an English teacher for many years, and the first woman city councilperson in her town of Oakwood, Ohio. When my grandfather retired, they moved to western Massachusetts to live down the road from my parents. During the summers I'd bike over to her apartment. She frowned at us when we put our elbows on the table at dinner, but always had a JELL-O and fruit dessert for us. In high school, when my sister gave me her ten-year-old hand-me-down car, Grandma paid the insurance. She also helped my sisters and me pay for college.

The last I had seen her was in Florida, just a short time before I biked into Mexico. Since by then she had lost a good deal of her mental capabilities, our last interactions were me reading to her. I was devastated that I wouldn't be at her service. Her encouragement and support are part of why I've been able to do so much with my life—I wanted so much to honor that fact.

Feeling the most alone of my entire journey, and not knowing what else to do, I remounted *del Fuego* the next day and continued south.

DAY 459
February 6—El Bolsón

Cold winds forced me to pull on my ripped and fraying leg warmers. My tattered GORE-TEX foot covers kept my toes warm as I pedaled in my Chaco sandals, my only pair of shoes. Their soles were peeling off, flapping at the edges.

The road would lead me to El Bolsón, a town of just over ten thousand people, which occupied the base of a deep, forested valley between serrated rock mountain ridges. Renowned for its artist culture, the town's economy is based on a combination of tourism and various crafts. Partially settled by hippies in the 1970s, El Bolsón had also declared itself "nuclear free," and had even thwarted governmental plans to establish a toxic waste dump 120 miles east of the town. But all of this was secondary to me. Of all the times I'd been in need of a friend, I was fortunate to have a friend here—Denali.

Denali and me at the outskirts of El Bolson

Denali, who'd grown up in a suburb of DC, had discovered and fallen in love with El Bolsón when studying abroad during his junior year of college. After he graduated he used some unexpected inheritance money—a few thousand dollars—to purchase a plot of land in the town with seven friends. Denali now shared this land communally. I'd heard he was building a small house on the property, and supported himself by making and selling wooden flutes.

Denali stood at the entrance to the town with a sign he made out of cardboard: DAVE! WELCOME TO EL BOLSÓN. His long, straight brown hair fell to his shoulders, and he sported a thick beard on his otherwise smooth young face.

"You hippie," I laughed, and gave him a hug. We'd joked in school that I'd ride to visit him in El Bolsón. I remembered sitting with him on the lawn in front of his house, circling EL BOLSÓN on the map, saying I'd someday make it there by bike. Now I had.

The next day the cold wind gave way to warm summer temperatures, demonstrating what Denali called the schizophrenic nature of Patagonian weather. In the warm sun I followed Denali to a large grassy plaza where, three times a week, over a hundred vendors sell artisan crafts, including wooden carvings, pottery, leather belts, and jewelry hand-crafted with colorful local stones.

Denali explained that for the first summer he'd sold flutes just at the craft fair, but now peddles them all over Argentina. I asked how he could afford to live.

"Well, I live really cheaply," he said. He needed only a few hundred dollars a month to survive there, and supplemented his artisan income with occasional carpentry and farming. "My life is great. I play music, and I make musical instruments so that others can play music." He then described his recent travels to music festivals.

I couldn't help but laugh. "I admire you," I told him, "but doesn't your college degree give you a sense of security others don't have? I mean, you have other options. If this fails, you can do something else. You don't *need* to support yourself through farming and making flutes." But, though this was true, Denali nonetheless isn't wealthy according to U.S. standards; as the son of elementary school teachers, he needed substantial financial aid to complete his college degree.

"You know, it's funny you say that. I no longer fear being poor, and it's not because of my degree. Since living here, I now know how to be a farmer, an electrician, a carpenter. And I'm quite good at it. Those are skills needed around the world." I wanted to heckle him, but I found myself feeling envious instead.

We spent the rest of the day wandering around the town, visiting various artisan booths. We stopped for a time to watch two of Denali's friends play the sitar and drums, producing surprisingly soothing discordant melodies and rhythms. At an art exhibit that evening I spent two hours staring at paintings of mythical lands, fantastic fictitious worlds on canvas—after which we joined the artist for dinner. I had been skeptical of this "town of artists," assuming it would be all show and no substance. Yet during our perambulations that day I found myself understanding why Denali lives in El Bolsón.

Denali's house, still under construction, was accessible only by foot, so the next day we hiked forty-five minutes over an evergreen hill and down to the banks of a turquoise river. The wooden frame two-story house had bales of hay serving as walls; these were covered with mud, creating a cement to keep the hay dry. Made of cheap and readily available materials, the thick hay walls would provide excellent insulation. Only one wall had been completed; a pile of hay bales lay under the roof, waiting to be erected.

"Other than the floor boards on the second floor and the hay bales," Denali said, "I built this house with my own hands, using materials found on this property."

"Can I lay one brick in your house?" I asked Denali.

"My house isn't made of brick."

"Even better. That way you will know which brick I laid."

After swimming in a river nearby and sitting on a rock in the sun, Denali and I borrowed a brick from an abandoned pile on the neighbor's property. Using mud as cement, I placed the brick flush with the bottom of one of the windowsills. "You see this brick," I proclaimed. "I laid this brick."

———————

I had little downtime on my journey. I spent my non-biking time visiting schools, publicizing the project, or updating the website. There were always pictures to upload, blog entries to edit, and email contacts to maintain, both on the road behind me and ahead of me. But in El Bolsón I took a rare few days neither biking nor working on Ride for Climate—instead I relaxed, taking in art exhibits and folk concerts. I found myself wanting to live in a community like this, where so many focus their energies on creative arts and everyone is within walking distance.

It was a great gift that I was able to spend this almost meditative time after the passing of my grandmother—and I felt fortunate to be with a friend and his community rather than on another isolated stretch of road. I took the opportunity one afternoon to write a eulogy for her, so at least my words could be at her service. I was sad that it was all I could do.

DAY 467
February 14—Patagonia, Tamed and Untamed

After a week in El Bolsón, I said goodbye to Denali and rode south through the steppe of Argentina. Patagonia, the name given for the southern end of the continent, is a cold and vast region the size of Alaska, Texas, and California combined, but with almost no people; if it were its own country, it would be among the least population-dense on Earth. Though the most rugged, it was also perhaps the most beautiful stretch of my journey. Patagonia is divided between Chile and Argentina by the Andes, with the wide dry Argentine steppe to the east and the fjords, glaciers, and temperate rain forests of Chile pressed up against the coastline to the west. The Argentinean plains are empty, with a harsh, almost constant wind that generally blows from northwest to southeast. Though my eventual goal, Tierra del Fuego, was south and slightly east of me, more often than not a substantial crosswind slowed my progress.

Every ten miles or so I passed the few wooden buildings of an estancia, an estate of a few hundred thousand acres. Such land, which is valued poorly, is used to raise tens of thousands of sheep and goats. Life on these plains is difficult, and little else seems to grow there—I saw few traces of animals other than the sheep droppings on the ground near where I set my tent. In the winter, the snow and constant wind make life especially hard. Many of those who work the estancias live like serfs, toiling long hours for little more than sheep meat in return. Not everyone gets to live in an artist colony.

As I rode this region, I felt emptier than I had during other stretches. As I had not visited as many schools in the previous few months, I wondered what my purpose was other than biking. I had tried to make the journey not just about me and my experiences, but to share it and connect it with students elsewhere. But in Patagonia I felt alone, unimportant, and far away from the rest of the world. My body responded by developing a slight cold, which meant biking fewer miles per day, and taking long naps on the roadside.

One night, camped in the wind shadow of a bush fifty feet from the empty dirt highway and miles from another soul, I thought of my grandmother and cried. I imagined the service I hadn't been able to attend, my family in an air-conditioned room in Florida, standing by her casket—without me there to hold them, offer support, read my eulogy. No one could hear my sobs.

As my tears subsided, lying in my sleeping bag, my mind wandered to the stars above me. If I make it to ninety years old, and if I lie on my death-bed while my grandchild travels the world, what will he or she see and hear and know while gazing up at the night sky? What will the world look like?

———————

The steppe was too lonely and monotonous for me, so at the next inter-section I encountered I headed west toward the Andes and back into Chile. The Andes in southern South America are less than half the height, by nearly 10,000 feet, of their counterparts far to the north. Although this meant I crossed the continental divide after minimal climbing, given that I was now farther from the equator than I'd ever been, I still passed by mountaintops covered by sheets of snow and ice.

On the Chilean side of the Andes, temperate evergreen rain forests lined deep glacier-carved valleys. Chile's famous dirt highway, the Car-retera Austral, led south, weaving a path through the mountains, fjords, glaciers, and lakes of southern Chile. With the aim of linking his coun-try with roads, the dictator Pinochet had ordered the road's construc-tion in the 1970s, but the southernmost portion wasn't finished until 2000. And the highway still doesn't lead all the way south—blocked by glaciers, Chile's southernmost cities are accessible only via Argentina. Pinochet's dream of connecting all of Chile is probably impossible.

I passed Lago General Carrera, South America's second-largest lake, which then emptied into the Río Baker, a flowing ribbon of deep turquoise between peaks with cascading glaciers. I rode in the middle of the empty dirt road. On occasion I passed by streams from which I filled my water bottles, drinking the cool water without treating it.

Chile's Carretera Austral

Javier by the Río Baker

Biking alone across vast open spaces, watching the scenery pan by as I counted miles, offers time to think and reflect. Much of this time is spent in an almost zen-like state, and much of the thinking is subconscious. But as I pedaled I also actively processed over a year of interactions and discoveries, trying to make sense of my varied experiences. I was also aware of how different it felt riding at this stage—with endless miles behind me, and few miles ahead—as compared to how it felt when the entirety of Latin America still lay ahead of me.

Around noon, alongside the giant Río Baker, I encountered another cyclist—a Chilean on a new Trek mountain bike who introduced himself as Javier. I asked where he was riding.

"To where they're going to dam the river," he replied.

"They're going to dam this river?"

Javier explained there were plans to build an enormous dam, and then clear a hundred-meter-wide swath of trees from which to run power lines—as well as pave the road we pedaled.

"How do you feel about that?"

"It makes me deeply sad. I bike to the place every day just to see it." Javier told me that he worked as a tour guide, leading tourists, mostly gringos, on hikes into these mountains. "Damming this river would kill the tourism."

There are many untamed rivers in South America where nations could build dams. According to the International Energy Agency, the continent could nearly quadruple the electricity it generates from hydropower. South America already gets nearly 70 percent of its electricity from such dams—far above the world average of about 20 percent—meaning that much of the region could increase its electricity use without using more fossil fuels. Also, because one can easily adjust the flow through a dam, hydropower works well in combination with intermittent sources such as wind and sun. On wind-free days, dams can run at full capacity; when the wind blows, water flow can be restricted.

But accomplishing this goal unfortunately involves the destruction of beautiful rivers like the Río Baker. And in Patagonia, as Javier had described, new high-tension power lines would cross hundreds of miles of wilderness in order to reach the cities in the north. I realized that I had arrived in a rare, fleeting moment: the moment after the dirt road was built, but before it was paved and the roadside developed. This moment would probably last only a decade or two. If I had arrived a few decades earlier, there would have been no road, and I would not have been able to visit this terrain by bike. And if I return in twenty years, I might find cities and buildings lining a paved highway instead of a beautifully remote dirt road. I savored that moment, but also realized it would not last.

Of course, even today there is no true wilderness. The wide steppe that I biked across is used to graze millions of sheep. And even along the Carretera Austral I occasionally passed plantation forests. Our reshaping of the natural world dates back to the first humans, who were so successful at hunting they forever altered the ecosystems by killing off creatures like the mastodons and woolly mammoths. Today, a full quarter of the Earth's land surface is used to graze livestock. Some scientists estimate that between 20 and 40 percent of all plant production on the planet is used by humans—from agriculture to ranch land to forestry. Environmentally, climate change is just part of the vast matrix of human activity that affects the quality of our life on this planet.

To provide basic (and hopefully better-than-basic) services for the nine or ten billion people expected to inhabit Earth later this century, we will have to continue to reshape the planet. For instance, we may need to plow the pristine salt flats of Bolivia in order to mine lithium for our electric cars, or we may need to dam more rivers in Patagonia to provide electricity for Chile. Or, as reported by a scientific paper in *Nature,* we may have to bring another 2.5 to 5 billion acres of land into cultivation—one to two times the area of the United States—in order to feed our civilization. Most of that new cultivated land will be in biodiverse regions of South America or Africa, potentially driving the extinction of countless flora and fauna. All of this change will likely happen *regardless* of whether we determine how to avoid global warming.

The Carretera Austral ended at yet another large lake, where a ferry took me across to Argentina. The ferry passed a glacier that had receded nine miles in the past century, creating a new, long freshwater fjord. To the west of the lake, out of view beyond the mountain peaks, was the Campo de Hielo Sur, the world's largest ice sheet outside Antarctica and Greenland—roughly the size of Connecticut. The Campo de Hielo Sur has been there for at least one hundred thousand years, if not for more than one million years. That ice is melting as well: a survey of the 270 glacial tongues that lead off it indicates that all but two are rapidly retreating.

Del Fuego and me on Lago General Carrera

The ferry dropped me off at a small Argentine customs office, accessible only by trail and boat. I pushed *del Fuego* for five miles along a trail through a forest, took another boat ferry across a small lake, and then continued on a road south. And I was rewarded for my efforts: the next day I passed by the majestic rock face of Mount Fitz Roy, whose silhouette is printed on every item made by the apparel company Patagonia. As no roads continued through the mountains—Pinochet's impossible road doesn't exist—I then biked east into the wide open terrain of the Argentine steppe, which extended away from the glaciated peaks and turquoise lakes. I rode yet another lonely week across this landscape, where rolling hills had been made smooth by past ice ages.

Fitz Roy Range

Steppe of Argentinean Patagonia

DAY 492
March 11—Punta Arenas

I crossed back into Chile to the city of Punta Arenas, population one hundred thousand, my last stop on the mainland of South America before taking a boat to Tierra del Fuego. Thanks to another contact found via my website—a young Chilean woman studying in Washington State—I'd been offered a place to stay: a three-room apartment where her mother lived with three children. My hostess told me, "The house is small, but the heart is big." I slept on the top bunk above the thirteen-year-old boy, who at first said little to me. I thought back to the first night I'd stayed with a family, just after I'd crossed the border into Mexico. Since then I had camped next to or slept inside the homes of eighty different people. The experiences had been fleeting, short, and many I'd met said as little as this boy. Yet I could vividly remember each one.

Students in Punta Arenas

Looking back on my journey, I almost felt the varied experiences as a great weight—as if I were physically carrying all my memories, somewhat in awe over how amazing it had been. I felt nostalgic for the journey though it wasn't even over yet, and this nostalgia mixed with the sadness and guilt I felt about my grandmother. These emotions felt somehow appropriate in the cold wind and gray skies of southern South America.

The Patagonian autumn was about to begin; the days were becoming shorter, and Chilean schools had just begun their academic year. Before leaving town, I visited one of the city's public schools, my last while biking in South America. It was refreshing to present my talk again, the students giving me one last push to complete the journey. I also appeared on the local Punta Arenas television news, which delighted the reticent boy who'd shared his bunk bed.

From Punta Arenas I boarded a ferry headed across the Strait of Magellan to the island of Tierra del Fuego. The boat motored east, with the wind at its back. The wind howled as whitecaps followed us across the twenty-mile-wide strait.

Wearing a worn-out fleece and a windbreaker that was no longer waterproof, I braved the cold wind and walked to the ferry's deck, pausing to look over *del Fuego*, who was as run-down and tattered as I was. The black paint on the sides of his aluminum crank arms, which connect to the pedals, had worn off from the countless times my sandals had rubbed against them. The bottom half of the front fender had broken off, and though it still bore the sticker of the Virgen de Guadalupe of Mexico, she was faded beyond recognition. Other stickers were equally pallid. The ONE PLANET ONE FUTURE sticker along the side now read —NET ONE FUTURE. The DEL FUEGO sticker on the top tube was now completely white. The gray handlebar bag on which I'd written in black letters TODOS SOMOS AMERICANOS—We are all Americans—had faded to a near uniform gray. I somewhat felt the way *del Fuego* looked—worn ragged by time and miles, and ready to reach the end of the road on the far side of the island of Tierra del Fuego.

NORTHERN PATAGONIA
February 4–March 17
Presentations: 2
Flat tires: 6
Miles: 1,316
Trip odometer: 15,466 miles

23 TIERRA DEL FUEGO
The End of the Road

DAY 499
March 18—The Winds of Patagonia

Although I might have been ready to reach the end of the road, I didn't know where that was. For so many months I had said I was riding to Tierra del Fuego, yet the two-hundred-mile-wide island shared by Argentina and Chile felt too big for me to claim victory by simply reaching it. I needed some ending point, some place where I could proclaim I had completed the full length of Latin America. I'd assumed the goal would be Ushuaia, Argentina's southernmost town and the city I planned to fly out of—but now that I approached it, I was reminded that Ushuaia was not the farthest south I could go. Should I cross the Beagle Channel to the Chilean town of Puerto Williams? Or venture to the series of islands leading to Cape Horn, an uninhabited piece of rock hammered by the waves and winds of the southern ocean? If I were a purist, perhaps I would take a boat to Cape Horn—that turning point for nineteenth-century gold seekers heading to California—and roll *del Fuego's* wheels across its tip.

Undecided, I pedaled south toward Ushuaia. After a night camped near the dirt road, I rejoined pavement the next afternoon. I soon reached a border crossing into Argentina, where a few small wooden buildings and a toll-booth-like structure stood on the otherwise windswept, empty plains of brown-green grass. In the equally empty customs building I walked straight up to the heavyset man behind the counter. He quickly flipped through the frayed pages of my passport.

"Your passport is full," he told me. "I'm sorry, you can't enter." Having crossed the Chilean-Argentine frontier seven times, the last five pages of my passport looked like a duel between stamp-wielding border agents.

"This page is empty," I said, pointing to the last page.

"This is a special page for addendums, not visas or stamps," he said without sympathy. My heart sank in realizing he might not let me through. The man then sighed and stared at the passport. No one else awaited his attention; there was no reason to hurry. We looked quietly at each other for twenty seconds of eternity as I considered the irony of the moment. Having successfully navigated sixteen countries, avoiding major theft, mishap, and complete mechanical failure, would I have my nearly complete journey cut short by this border agent?

Finally the man tilted his head, looked down, and imprinted a blurred stamp on that final page. "Go on," he said.

Reprieved, *del Fuego* and I continued south through a wasteland of rolling hills, the wind blowing furiously at my back. All vegetation was grazed short, although the sheep responsible were somehow never to be seen, phantom livestock roaming the land. The wind at my back gradually grew in strength as I rode, reducing the effort of each pedal stroke.

Patagonia is one of the windiest regions of the world, a condition resulting from its particular geography. The tip of South America juts farther south than any other land mass: almost 2,500 more miles than Africa, 1,200 miles more than Australia, and 600 miles more than New Zealand. If you were to sail from the east coast of Patagonia into the Atlantic, continuing east with the wind at your back, keeping to the 50th parallel south, the next land you'd hit would be the west coast of Patagonia. Thus, with no land mass to slow it, the prevailing west-to-east wind builds in strength as it circles the southern hemisphere. The result is the strong, sustained winds of the region, which average over twenty miles per hour—and which hold an enormous potential for renewable power. One study estimated that the installation of turbines across Argentina could produce, on average, over 4.5 terawatts of electricity—about twice as much electricity as the entire world currently produces. Of course,

Patagonia is a far distance from the major consumers of electricity, so we'd need to also store or transport the energy. Nevertheless, the potential is there.

Fortunately, *del Fuego* and I had no trouble harnessing the power of this wind, and as the afternoon progressed the air currents beating against my back grew even stronger. For the first time in my life I experienced a tailwind that blew faster than I could ride. We rolled at twenty-five miles per hour—without pedaling—my jacket and pants shaking loudly, the Argentinean flag on my rear rack blowing forward as if I were riding backward. I stopped briefly but almost couldn't stand upright, and didn't dare open a pannier lest I lose its contents to the air.

After fifty miles of cold-wind-assisted effort-free riding, I approached Río Grande, a city of about fifty thousand people two days away from Ushuaia. The city had short buildings and wide streets with numerous street lamps to provide light during the long winter. I asked for directions to the fire station.

The station was built like a giant hangar; inside it felt like a bomb shelter, completely protected from the wind. Unlike my last such attempt, this fire chief enthusiastically invited me to stay the night, telling me they often hosted travelers. Tierra del Fuego serves as the end or beginning of many cross-continent journeys.

Entries in three notebook-size guest books in the lounge were testament to these numerous overnight stays. Some travelers simply crossed Patagonia—others had crossed the entire world. One cyclist had come from San Diego, another from Canada. One traveler, starting from Alaska, had been on the road for nineteen months; two German carpenters had walked for three years. A French cyclist on a global tour beat them all.

I wanted to meet each traveler and hug them: You did it! You made it! And soon, so would I.

The following day I headed toward the mountains rising up from the land fifty miles away. The wind continued to howl, but it blew from my right to my left, making progress more difficult. In time the open grassland gave way to forests of twenty-foot-tall trees, their small leaves hinting red and yellow, signaling the coming autumn. Somewhat protected from the wind, I camped that night hidden in a forest, feeling sorry I had so few remaining nights on this grand journey.

Come morning, I crossed a pass between mountains and biked alongside the long U-shaped valley leading to the southern coast and the city of Ushuaia.

Bomberos in Río Grande

Descending toward the city of Ushuaia

DAY 502
March 21—Route J

The city of Ushuaia curved around a bay on the Beagle Channel, a glacially carved valley that dipped below sea level, thus dividing Tierra del Fuego from its neighboring island to the south. The channel was named after the HMS *Beagle,* the first European ship to survey its waters and later the vessel carrying Darwin on his travels. When I arrived I saw a large ship floating in the water; at the edge of town sat shipping containers just like those I'd seen in Panama and Manaus and Los Angeles. Ushuaia is proud of its shipping port status. Deeming Chile's Puerto Williams, on the island to the south, just "a town," Ushuaia claims it's the world's southernmost city. A sign welcomed me at its entrance, proclaiming exactly that.

During the last ice age, Ushuaia, the Beagle Chanel, and nearly everything else I saw sat beneath a mile of ice. That was when the Earth was about 10 degrees Fahrenheit colder than it is today. If today's Earth were to warm another 10 degrees, would this landscape be unrecognizable? Would crocodiles swim in the widened Beagle Channel? The last time the Earth was that warm, about fifty-five million years ago, such reptiles survived in the Arctic Ocean near the North Pole.

The answer is we don't know. While the Arctic is expected to warm much more than the rest of the planet, the southern tip of South America is not. The reason is that, unlike the Arctic, which is surrounded by land, the southern hemisphere is mostly ocean—which will buffer the region against excessive warming. Nonetheless, though it likely won't warm as much as will Alaska or Greenland, temperatures will still rise, greatly affecting regional vegetation. Much as the cold-seeking plants at tops of mountains run out of mountain as the climate warms, those at the tip of the continent have no land farther south to migrate to. If the temperature continues to rise, at some point they perish.

———————

Many cycle tourists end their bike trips in Ushuaia. Others continue to the end of Route 3, which curves about twelve miles out of town, ending in Tierra del Fuego National Park. I followed that road to its culmination, but it didn't feel like the end of the journey: a sign marking the ultimate cul-de-sac was overrun by tourists eager to see the end of the world. I had expected an anticlimactic ending, but I had hoped it would be more authentic, more personal. It was upsetting to pedal for months on end to reach a place that felt packaged for a photo op, another "Picture Spot" at Disney World. Wanting to celebrate my achievement, I walked a distance away; when I thought no one was watching I stripped naked and leapt feet first into the 50-degree waters of the Beagle Channel. When I soon reemerged I realized a large group of elderly Argentine tourists had watched me from a boardwalk above. A few clapped.

Entrance to Ushuaia

I needed a different ending. That night, back at a hostel in Ushuaia, among people of many different nationalities all speaking in English, I again pored over my map. It showed a Route J, a squiggly dirt road leading southeast from Ushuaia, following the coast for about sixty miles before ending at what the map suggested was a washed-out bridge. And because the southeast corner of Tierra del Fuego juts out like a half-pointed toe, Route J ended at a point farther south than any road on Tierra del Fuego—even slightly farther south than any road on Isla Navarino, the Chilean island just across the Beagle Channel. This could be the southernmost road in the Americas. I asked local officials and a tourism agent what I'd find at the end of this road: they didn't know. Perfect.

The following afternoon I pedaled onto Route J. Pavement gave way to packed dirt; a truck or van bounced past me only once every half hour or so. As sunlight faded behind thick clouds, I pulled off the road and set up my last campsite in Latin America. I placed my tent on a moist meadow in the forest, between short trees with wide gray trunks and leaves kissed with gold and yellow. After cooking a pasta dinner, I crawled into my tent.

I soon heard the sound of hooves trotting across wet ground, approaching my tent. Sticking my head out, I was unsurprised to find two boys on horseback; nor was I surprised by the same questions I'd answered a million times—Where are you from? How long have you been traveling? Do they pay you for this? When being asked such questions I always felt I was harboring a secret, never revealing the essence of what they wanted to know—*what it was really like to travel this way for so long*. But no one requested that specific answer, and anyway, I would have been incapable of explaining it.

After warning me that bulls on their land could be aggressive, the boys trotted off, leaving me to sleep alone in the semi-wilderness. Cool air flowed through the mesh opening at the bottom of the tent, and I fell asleep to the sound of a nearby stream and the rustling of leaves.

In the morning I encountered no bulls. After a breakfast of corn flakes and powdered milk in my titanium pot, I broke camp, loaded my saddlebags, and pedaled south along Route J. According to the map, the route would end at a washed-out bridge over the Río Moat. It felt odd to think that all those months of biking would culminate in a washed-out bridge.

My wheels bounced over the rocks in the dirt road. The leaves of stunted trees held subtle highlights of red and gold, and the sun, charting a low path across the sky, offered the angled light of fall. Only two cars passed me. For the final twenty miles the road paralleled the four-mile-wide Beagle Channel, which, aside from the very occasional building, must have closely resembled the passage seen by Darwin almost two hundred years earlier. When I reached the Río Moat, a river a mere twenty feet across, I found a new red steel bridge across the water—and yet more road. I continued, nervous with anticipation.

The road climbed, ending after half a mile at a white, one-story building on a bluff overlooking the choppy waters of the Beagle Channel. Tall antennas protruded from the building's turquoise roof, and three wiry dogs barked over the hum of a generator. The dogs ran up to me, two stopping short, wagging their tails, the third approaching and sniffing my saddlebags.

I dismounted, walked to the door, and knocked.

A man in khaki pants and brown sweatshirt opened the door. "Hello," he said.

"Ahh . . . hi. I biked here from California."

"Really?"

He looked at me and tilted his head. I guessed he didn't receive many visitors.

"What is this building?"

"It's the Prefectura Cabo Segundo. We work for the Coast Guard."

"Oh—you monitor ships as they enter the Beagle Channel?"

"Exactly. Come in," he beckoned.

The man introduced himself as César. He worked with Juan and Eduardo, who sat at desks that looked out a window toward the water. No ships were in the channel. They wore sweatshirts and work boots, their chairs angled for watching a news program via DirectTV.

"This is the end of the southernmost road in South America," I told César.

"No, it isn't. Puerto Williams has roads, and it's farther south," he replied, matter-of-fact.

"No, we're farther south," I countered. "Here, look on this map." We walked over to a huge nautical map on their wall depicting the southeastern portion of Tierra del Fuego. The map showed no roads, but it did indicate the location of settlements. The blue ocean and waterways were covered with numbers showing the water's depth.

César studied the map carefully, comparing where we were to the Chilean town on the next island. "You're right," he said without inflection. The eastern part of Tierra del Fuego curved farther south than Puerto Williams. "This is the farthest-south road." This simple proclamation felt more final than the road sign near Ushuaia, perhaps partly because those actually at the end of the road hadn't thought they were at the end of the road. For that matter, the people I'd met all along my travels hadn't seen themselves as on the road to Tierra del Fuego. The journey was my construct, not theirs.

I walked outside to where a grassy overlook gave a commanding view across the channel. Dark, textured clouds hung overhead, mirroring the water below. The clouds were thinner in the distance, and the horizon glowed with a yellow-white lining. Only the dirt road disappearing to the northwest showed any sign of human settlement.

View from the end of Route J

I walked back inside, where Juan, César, and Eduardo offered me leftover steak and potatoes. César and Eduardo then leaned back in their chairs to watch a movie while Juan manned the radio.

I almost laughed when I realized they were watching *Waterworld*, an overbudget Kevin Costner film that takes place in a futuristic dystopia where "environmental destruction melted the ice caps," somehow covering all the world's land with ocean. Warring bands of humans survive on boats and floating cities, and the plot hinges around the custody of a young girl, who has tattooed on her back a map to "dry land." I joined the men to watch the film, not sure whether to tell them I had spent the last year and a half traveling and talking about the environmental destruction that could lead to sea-level rise and civil unrest. I cut another piece of beef instead.

I thought then, as I think more strongly now, that this movie conveys a narrative of environmentalism we should avoid telling. The movie essentially says: if we don't correct our ways, the result will be the elimination of life as we know it. Such tales can lead to fear and inaction. Moreover, the film suggests the only solution to our environmental problems is to revert to the Stone Age. We should instead tell a different story: that a better, wealthier future is possible, where we solve poverty *and* climate change. Few Hollywood movies embrace that idea.

In the building at the end of Tierra del Fuego, I watched to the end of the movie. And while the environmental narrative didn't speak to me, the ending did. After much fighting and many explosions, Costner's character saved the girl, and his friend deciphered the map on her back. He and a small clan then charted a course across a near-endless ocean, eventually reaching an empty tropical island. Celebrating their arrival, they exulted at sights never seen before: a fresh-water stream, birds perched on trees, a mountain rising above the forest. They had succeeded.

Standing on land for the first time in his life, having vanquished his enemies and reached what they considered an impossible dream, Costner looked at his friend and said, "It doesn't move right."

His friend, a skinny man with wild gray hair, replied, "That's land sickness. It'll take some getting used to."

"I don't belong here," Costner proclaimed. The movie ends with him sailing on his small raft into the open sea.

César changed the channel to the national news out of Buenos Aires. I washed my plate. A few minutes later a white pickup with oversized wheels pulled up in the driveway. César, Juan, and Eduardo unloaded its cargo, resupplying the station with barrels of gasoline, drinking water, and various essentials. The driver, an older man with twisted gray hair, offered me a lift. I looked across the Beagle Channel, a view I'd pedaled almost sixteen thousand miles for, then accepted the offer, loading *del Fuego* and myself onto the back of the truck.

Bouncing on the dirt road, feeling the breeze through my greasy hair, I felt disoriented and carsick. In two days, I would begin the series of flights that would take me back to the United States, undoing in less than twenty-four hours by air what it had taken seventeen months to accomplish by bicycle.

TIERRA DEL FUEGO
March 18–23
Presentations: 0
Flat tires: 1
Miles: 399
Trip odometer: 15,865 miles

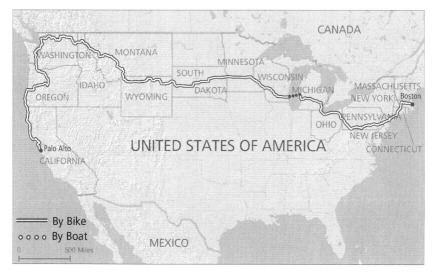

Ride for Climate USA route

EPILOGUE ⚲
Ride for Climate USA

United States population: 314 million
Annual per capita GDP: $50,000
Annual per capita CO$_2$ emissions from fossil fuels: 17.6 tons

Though I'd finally reached the tip of South America, my trip wasn't over. I flew to the East Coast of the United States and then began biking to California with Bill Bradlee, the climate advocate who co-organized the second leg of Ride for Climate. The U.S. trip, a five-month, five-thousand-mile journey, was a far different adventure from the one across Latin America. And yet, by crossing my own country, I better understood both the lessons from Latin America and how my two years on the road had changed me.

I realize the apparent lunacy of this second trip. After reaching my goal of Tierra del Fuego, having biked sixteen thousand miles to do so, how could I possibly want to ride for another several months? Emotionally, I was conflicted about this decision. A part of me wanted to stop biking: to live in one place and see friends on a regular basis—but I'd also become addicted to the exertion of biking for hours every day, as well as to the constantly changing landscapes of those hours. The real reason I continued, though, was not because I wanted more adventure; it was because I wanted to make a difference in my own country.

But my commitment didn't stop me from hesitating before continuing. In early April, just a week after reaching Tierra del Fuego, I found myself sitting in my parents' house in Massachusetts. I remember feeling disoriented, trying to figure out what had just happened to me. Only a pile of *bombero* uniforms and eighty gigabytes of photos convinced me I had, in fact, just biked to the far end of the world.

But I didn't have much time to ponder the journey I'd just completed—I had a new journey to prepare for. After barely enough time to catch my breath and replace or repair numerous parts on *del Fuego*, I set out again with my new companion.

Though we'd spoken by phone numerous times, I met Bill only a week before departing. A decade older than I was, he was as I expected: soft-spoken, intelligent, cautious, and organized, with a quiet but biting wit. He worked at an organization that engaged communities of faith on climate change. While I'd been biking across Latin America, he had, in his spare time, spent months researching our route, setting up talks, and planning logistics. Then, once I joined him, Bill quit his job for the trip, grateful for the chance to engage the public on climate change in a new way. We set out on April 21.

———

Publicizing Ride for Climate in the States was a far cry from doing so in Latin America. I'd previously often arranged a presentation and given my talk on the same day, but in the U.S. our presentations at community centers or schools were planned about a month in advance. In order to keep to those commitments we had to develop a detailed itinerary, one that specified, several weeks out, the miles we needed to cover each day. Having a schedule by necessity reduced the sense of adventure, but it vastly improved outreach—enabling us to address entire school assemblies rather than just individual classrooms.

It was harder, though, to get press coverage in the United States. Not being a foreigner, I was certainly less of a novelty. And probably the media didn't consider the trip as grand, whereas it certainly had been in Latin America. So ultimately we appeared in only one nationally broadcast piece, which, interestingly, was for Spanish-speaking television. Nonetheless, we did get a few dozen stories in local print, radio, and television.

Bill speaking at Maumee Valley Country Day School, Toledo, Ohio

What struck me most about the trip were the contrasts between Latin America and the United States, contrasts that brought my experiences of the previous months into greater focus. For one, I surprisingly felt less safe biking across the U.S.—simply because our roads, having been engineered for cars, leave little room for pedestrians and bicycles. Plus, U.S. motorists often aren't mentally prepared for cyclists—or sometimes even pedestrians. By comparison, even though motorists in Latin America are less likely to follow traffic laws, they are more accustomed to seeing cyclists, pedestrians, even horses on the road. They are prepared to slow down and swerve out of the way—and are even more likely to give a friendly wave.

The most obvious contrast I experienced on this second leg was how incredibly wealthy we North Americans are. Our houses are enormous, our cars luxurious. It felt strange to bike through newly built suburbs, riding past three-story, three-car-garage houses with freshly cut grass. I couldn't help comparing these homes in my mind's eye to the adobe huts of Honduras or Peru, modest structures one-tenth their size.

The thing that saddened me the most about these neighborhoods was they mostly lacked people to talk to. In many communities in Latin America I'd often encountered people sitting outside their homes, and I had felt welcome to approach them for directions or a place to camp.

But in the U.S. suburbs, people drive straight into their garages, entering their houses without interacting with anyone. Bill and I saw few Americans sitting on their front porches, which made it harder to ask for a place to camp. So instead we tried approaching people at gas stations, where we quickly found individuals willing to let us pitch a tent on their lawn.

Although I disliked many of these communities, I also recognized how much more agreeable they were than most I'd seen on my trip. The large houses in particular brought to mind Latin Americans who'd said they wanted to live in homes like they'd seen on television. However wasteful big homes may be—and some have huge environmental footprints—one can't deny that their comfort trumps a makeshift structure in a mega-city slum.

Suburb in eastern Pennsylvania

But for all the differences between Latin America and the States, there was also a strong similarity: the generosity of the people we encountered. In Wyoming we stayed at a rancher's farm and enjoyed its fresh beef. In Montana two strangers even offered the use of their homes while they were away, trusting us with their spare keys. In Ohio we turned down a meal almost every day, a place to stay every other night. The kind offers were so profuse that Bill and I mailed a pile of thank-you notes at the end of our journey.

And, while Latin America had shown me many places subject to the effects of climate change, some of the most blatant impacts I saw were in my own country. We stayed with farmers in several states who complained that drought greatly reduced their yields of wheat or corn. We saw how Wyoming's Bighorn Mountains have lost miles and miles of ponderosa pine forest to bark beetles—infestations linked to warmer temperatures and drier weather. And we rode by communities in Connecticut where the coastline butts up against expensive homes, water nearly lapping against the doors. But it isn't just upscale real estate that's at risk: about four million Americans live so close to the ocean that the projected rise of sea level, perhaps by three feet this century, will endanger their homes and communities.

Wyoming farmer who complained about drought

During our five months crossing the U.S. we spoke about climate change at about fifty talks addressing a few thousand people, and reached many more through our website and the media. I'm proud of this outreach. We learned later that we'd inspired a number of the people we'd reached to think more about their energy consumption. Some switched to more efficient light bulbs; a few now biked to work. One even helped lead a campaign against a coal power plant in her community. While she might have led that campaign anyway, I'd like to think our project encouraged her to speak out.

Of course, most of the people we reached were already aware that climate change is a growing concern, and were open to learning what we had to share. Those who don't consider climate change a serious issue are harder to reach, and were less likely to attend our talks. But whenever we got a chance—at gas stations, in cafés, along the road-side—we tried to engage strangers in conversation about the issue.

Waterfront homes in Connecticut

Some of these conversations were frustrating. A few people, like the farmer I'd met in California's Central Valley, disputed the science. We met a beekeeper in Pennsylvania who complained that the changing weather and warmer winters were harming his bees—but he also declared there "was no way to prove" that humans caused global warming and argued against taking action. An owner of a gas station in South Dakota said flatly: "We're conservative, so we don't believe in that," as if climate change were a choice of faith that only liberal people should believe in. But most we met weren't confrontational—they just didn't feel strongly one way or the other. And our anecdotal findings mirrored national public opinion polls: although many Americans may want action on climate change, a majority of Americans list global warming near the bottom of national priorities. When given a list of twenty different issues—ranging from deficit reduction to Medicare to tax policy—U.S. citizens consistently ranked global warming as the least important.

Surprisingly, the average person in Latin America is more worried about climate change than is the average person in the United States. According to several Gallup polls asked of people who said they knew about climate change, as of 2012 more than two-thirds of Latin Americans said it was a "very serious" threat to them or their families, a concern shared by less than one-fifth of respondents in the U.S. And it isn't just Latin America—among people who know about climate change, public concern is higher in *most countries of the world* than it is in the United States.

This deeply saddens me. As climate change affects our entire planet, every country needs to play a role in solving our collective problem. But, though we must work together, the United States has a special role to play. I believed this when I began my trip, and I still believe it today. But over the course of my journey my reason for this belief changed. I'd initially considered the importance of reducing emissions in terms of guilt-driven responsibility: as the United States is historically the largest polluter, it is the United States who should largely cut back on its pollution.

Now I see the issue differently. My views are now more forward-looking, and they concern people more than they do carbon. I believe it is not enough for the U.S. to reduce its emissions; any reductions can and probably will be matched by increased emissions from the developing world. But reducing our emissions while demanding that others not develop as we have is not an option either. More access to energy is a good thing—it means many families like those I stayed with in Latin America would get electricity in their homes; it means countless people could climb out of poverty. And we can't criticize poorer nations for using fossil fuels if there aren't cheap alternatives. I think the only solution to our global concern is to develop energy that is both clean and cost-competitive.

But achieving that will cost money. It will take massive investment to figure out how to revolutionize energy. Luckily, players across the world are already contributing. Countries like Brazil and China are currently working to make clean energy cheaper, investing in renewable fuels and developing affordable solar panels. But it's the countries with the most economic resources and research capacity—like the United States—who simply have the greatest ability to make this happen.

But the roles of various countries—and, more specifically, the role of my own country—is a much longer discussion, one I couldn't address by taking a long bike trip. My goal was to raise awareness, to get people to care about this issue. Meaningful policy change can only result from popular support.

One thing I've realized from this trip: it's difficult to get people who live local lives to understand global problems—and most of us, understandably, live very local lives. We spend most of our time in the town where we live, and often our most meaningful interactions are with only our fellow residents. But our world is becoming more connected, and every day we are more likely to interact, at least indirectly, with people from around the globe. We buy fruit from Central and South America, electronics from Asia, oil from the Middle East, cars from Europe. And when we drive European cars fueled by Middle Eastern oil, we change the atmosphere for everyone on the planet—whether they benefited from the transaction or not.

Yet while many are aware of these global connections—say, when they remove the GROWN IN CHILE sticker from an avocado, or peel a banana imported from Central America—few have a concrete understanding of what these foreign countries are like, and even fewer know people who live there. Only about one out of every three Americans has a passport, so most U.S. citizens have never traveled beyond our borders.

I was very aware of this "global disconnect" while biking across the U.S. When I'd mention to someone I'd just biked to the tip of South America, I often felt I might as well have said I'd spent the last year on the Moon. For example, I was allowed to review the essays a class of college freshmen wrote about my presentation. A surprising number of students said they didn't know much about global poverty; one wrote: "Living in the United States, I never realized that a lot of the world doesn't have the same lifestyle as we do."

———

I stated earlier on that when I set out on my Latin American journey I had two things in mind: one, to have an adventure by bicycle; and two, to raise awareness about climate change in the process. In retrospect, those goals were actually secondary to the primary benefit of my trip—which was that the people I met throughout the sixteen countries I traversed changed the way I see the world. It's difficult to compare an earlier self with a later self—unaware of how our subconscious has gradually changed, we can forget how we used to interpret the world around us. But I definitely feel more connected to the people with whom we share this planet. Thanks to my travels, I now have a mental image of what people's lives can be like across the Americas. Over the course of two years I had the privilege of staying in or near the homes of nearly one hundred people, sleeping at almost forty fire stations, and sharing meals with more interesting and diverse individuals than I could possibly count.

The person most converted on a mission is usually the missionary, and I'm no exception. I care more about climate change than I did when I started, but I also think more often about my place in global society. The experiences of my trip have stayed with me such that, when people ask what they can do about climate change, one of my suggestions is simple: travel. Yes, travel can be carbon intensive, but if we want to address global problems, we need to build connections with people across the globe—we need to understand what life is like for people unlike ourselves, and to visit the places at risk. The solution to climate change isn't to stop traveling via polluting car or polluting plane; it is to develop technology whereby cars and planes don't produce carbon dioxide, and whereby we can become even more globally connected.

When I think about how lucky I was to have this adventure, I think first about the people, and about how much I learned from them. After a hard day of riding, nothing felt better than to have complete strangers offer a meal or a roof—and then also share with me a bit of their lives in their corner of the world. I don't claim the people I met were entirely representative of the Latin American populace—I talked to far more men than women, as was culturally more appropriate for a single man, and I didn't stay with the urban poor, who make up a large portion of the population. But those I met nonetheless formed a diverse group, and each of them taught me a great deal.

There was Carleton, the former mayoral candidate in Tegucigalpa who gave me a tour of town in his pickup; Melvin and Rosa, the subsistence farmers in rural Honduras who shared their dinner of rice and beans in a house with dirt floors; Jorge, the exiled former president of

Guatemala, with an office near the top of a Panama City skyscraper; César, the drunken oil-rig worker who lived by Lake Maracaibo in a wooden hut; Juan and Victoria, the petroleum engineers who also lived near lake Maracaibo, but in a high-rise apartment; Dutchman, the retired American with the jaguar in Belize; Ivan, my guide into the Andes who expected me to know kung fu; the beggar children in Chiapas, Mexico, who taught me how to say "you are pretty" in Tzeltal; Ricardo, the bicycle advocate in Bogotá who ran a small nonprofit; Baltazar, the shrimp farmer in Oaxaca, Mexico, who beamed when showing off his small catch; Jerry, the fire chief in Juanjuí, who watched the rain put out a fire in the Peruvian jungle; Pedro, the grandfather on the beach in central Chile; Denali, my flute-making friend in Patagonia; and countless more, including the people we stayed with while crossing the United States. All of these individuals share the planet with us—and all of them live just a bike ride away.

UNITED STATES
April–September 2007
Presentations: 44
Flat tires: 17
Miles: 5,138
Trip odometer: 21,003 miles

RIDE FOR CLIMATE
November 2005 to September 2007
Total presentations: 111
Total flat tires: 66
Trip odometer: 21,003 miles

APPENDIX I
What You Can Do

Each of us has incredible potential to make a difference in the world. Think of any great accomplishment in history, whether technological, scientific, or political: ultimately it was individuals or groups of individuals who effected that change. Simply put, if you think you as an individual can't make a meaningful difference, history proves otherwise.

With that in mind, we have to ask two questions. One: if we lived in a world in which we'd successfully addressed climate change, what would that world look like? And two: what can you or I do to help create that world?

I believe a successful future looks like this: the billions of people who live involuntarily "off the grid" would gain affordable access to electricity, from which they could enjoy a significantly improved quality of life. Now, as those who currently use almost no energy gain basic—or hopefully much better than basic—services, humanity's consumption of energy would significantly increase—even if some of us are able to (as we should) conserve our energy use. But, even with that increase, carbon dioxide emissions would fall—as the result of affordable clean energy having replaced fossil fuels. In brief, this future entails humanity figuring out how to live comfortably without polluting.

Now, some may say this is a pie-in-the-sky vision, but it is possible. And getting there is somewhat straightforward: we need a massive global investment in clean energy, combined with intelligent curtailing of fossil fuels. According to estimates by the World Economic Forum and the Climate Policy Initiative, if we want to limit the increase in average temperature to 2 degrees Celsius (3.5 degrees Fahrenheit), this investment should be around $5 trillion a year—which is roughly the GDP of Japan. Yet, to date, investment has reached only one-tenth this level.

Two degrees Celsius is the maximum warming considered "safe" by the international community involved in negotiating climate treaties. But most experts believe we will miss this goal—and that we are currently on track to warm the planet by 3 or 4 degrees Celsius. That assessment doesn't, however, mean we should give up. The fact that we are on pace to pass 2 degrees doesn't make our efforts less important; it makes them more important.

Each of us plays a role in shaping the future of humankind; that role begins in deciding which course we take every day. Each of us has an effect on our environment, and we each have control over what we personally contribute. Even more important: each of us has a vote.

We certainly can't change the world in a day—but we've got to start somewhere. Below are three of the most effective ways you can reduce climate change, in order of importance. Even a small change now can make a big difference; hopefully you can follow up your success with larger steps. And, while these suggestions are tailored to a U.S. audience—and they are just my suggestions—the general ideas apply for almost anywhere in the world.

1. VOTE, CONTACT YOUR REPRESENTATIVES, AND GET INVOLVED

In short, to address climate change, governments must set the rules. We need laws that promote investment in climate solutions. This could take the form of a tax on carbon dioxide emissions, an approach that not only encourages polluters to reduce emissions but also penalizes those who don't. Another method is the "cap-and-trade" system, which puts an economy-wide limit, or "cap," on emissions; this approach allows polluters to determine themselves how best to meet this goal. Many other useful regulations, such as mandates to increase the efficiency of buildings and electronics, can meaningfully reduce emissions. All these policies would help us make clean energy cheaper by forcing us to develop innovative technologies. Personally, I'd also like to see a major increase in government-funded research and development so as to boost our progress toward this goal.

But the specifics of policies such as these won't matter much if there isn't demand for them. Only with demand are we likely to get meaningful legislation. That's where you come in.

VOTE. Vote for candidates who support climate action; don't vote for those who don't. And feel free to tell those you voted for why they won your vote—and tell those you declined *why* they lost yours.

CONTACT YOUR REPRESENTATIVES. Let your representatives know you want climate action. Congressional staffers in Washington are paid to sit at desks and track how many calls they receive on various issues: add your opinion to the tally. You don't even need to have a full spiel ready—they don't have time to do more than register your preference. It's really very easy: just go to usa.gov, select CONTACT GOVERNMENT, and then select ELECTED OFFICIALS to locate the contact information for your representatives. Tell them you want action on climate change—and that their action or lack thereof will determine your vote in the next election.

And while it's true national legislation has the greatest impact overall, the policies of state and local governments also contribute meaningfully to the issue. A number of different laws and regulations effected by a handful of states, mostly along the Pacific coast or in the northeast, are making significant progress to reduce emissions. A number of states have also adopted "Renewable Portfolio Standards," which mandate that the state produce a minimum amount of electricity from renewable sources. Your support of state-level actions like these and others could ultimately have a large effect on reducing emissions.

You can also make an enormous difference at a local level, even by simply showing up to town hall meetings. Many cities have "climate action plans"—does yours? Find out, and find out if there is some way to support that action. Does your city have plans to purchase cleaner electricity, or to build more bike lanes to make alternative transportation safer and more enjoyable? Such efforts need popular support; express yours.

JOIN GROUPS THAT ARE PUSHING FOR SUCH CHANGE. This is one of the most important things you can do. Why? Because on the national scale climate change is barely on the political radar. It received scarce mention during the 2012 presidential elections, and only a few candidates make it a significant issue in their electoral platforms. You can help change this reality.

Find out what member-based organizations speak most for you. Even such groups as the American Lung Association and the NAACP support climate legislation. Don't just speak and act alone; join with others to amplify your voice.

JOIN CLIMATE RIDE. If you want to ride a bike and make a difference on climate change, you are in luck. Climate Ride (www.climateride.org—not to be confused with my site, www.rideforclimate.org) offers week-long bike tours where you can raise money for your favorite environmental NGO (non-governmental organization), whether it is350.org, NRDC, the Sierra Club, your local bicycle advocacy organization, or any other group you care about that is making a difference. I participated in Climate Ride's East and West Coast rides, and both are fantastic ways to both raise money and get involved with other people seeking change. Obviously (as I think I've demonstrated), a single bike ride isn't going to solve climate change, but it can make a difference.

2. SHARE YOUR VIEWS: TELL YOUR FRIENDS

None of us can stave off the threat of climate change on our own; it will take collective effort to fully address this global concern. And while it might be challenging to convince a complete stranger of the peril to our planet, you have a much better chance of getting your friends and family to consider the issue more seriously. Help them understand that climate change will affect all of us—and that we need to act now.

But first: equip yourself with the facts. A small but vocal percent of the U.S. public denies the existence of climate change. Though their arguments are unsupported by science, they can nonetheless confuse people into maintaining the status quo. This makes it all the more important to refute such claims.

Grist.org has a great series on how to talk to "climate skeptics." An even more comprehensive, regularly updated list of talking points can be found at skepticalscience.com, which also offers iPhone and Android apps that provide responses to almost any "arguments" put forth by climate skeptics.

3. REDUCE YOUR EMISSIONS

To avoid the worst of climate change, we must develop the means to live comfortable, even prosperous lives without polluting. As stated above, the most efficient route to this goal is likely through investment, technology, and policy: large-scale investment in clean energy technologies, and government policies effectively curtailing the use of fossil fuels.

But the need for large-scale policy advances doesn't change the fact that you as an individual can make small improvements every day. Part of "figuring out" how to live comfortably without polluting involves each of us regularly taking steps toward this goal. As it happens, if you live in the United States you can likely reduce your emissions and improve your quality of life simultaneously, especially if you work to improve the efficiency of your transportation and your home. Simple, unexciting efforts such as replacing lighting, improving insulation, and upgrading appliances can save many people money and reduce emissions. Another example: as much as many Americans love eating meat, nutritionists advise we'd live longer, healthier lives if we ate less meat, especially red meat. But that dietary improvement would reap another benefit: reduced greenhouse gases.

Okay, so where do we start? First: calculate your personal impact on global warming. My favorite web-based carbon calculator, www .coolcalifornia.org, does more than just indicate your carbon footprint—it can identify not only which specific changes would lower your impact, but also how much money such changes would save. You can also compare your emissions to those of the average person in your U.S. zip code. Another website, www.oroeco.com, also lets you compare your emissions with your peers—and earn rewards for reducing them.

———

We have another consideration to add to all of the above: the cost of global warming. The good news is that the sooner we address climate change the cheaper the total cost will be. It's really very simple: we will save money by taking action now. Economists estimate that avoiding the worst of global warming—which means acting now—will be relatively inexpensive, costing us only around one percent or less of our total economy. It won't come free, and not without major effort—but that's a small price to pay to safeguard the global environment against a hot, unstable, and uncertain future. Imagine if we each made a one-percent effort—that is all we would need. I believe we are each capable of contributing our one-percent effort against climate change.

———

Finally, it's important that we all take the long view. We won't "succeed" overnight: it will take decades for us to cut sufficient greenhouse gas pollution—and to adapt to the consequences of climate change already set in motion. This is an ultra-marathon, not a fifty-yard dash. And while we shouldn't expect to "solve" the problem in any given year, we should expect to continually move closer toward the goal of a clean, healthy, prosperous future. To do that, we need your help: your help taking simple steps, your help building broad public support for larger action. The 7 billion people who call Earth home are counting on it.

APPENDIX II
General Recommendations on Bike Touring

I can't tell you how you should bike tour, because there is no right way—there are advantages to each approach. Some people tow trailers, some have panniers. Some people travel so light they saw off their toothbrushes; others would strap a guitar to their rear pannier and carry a dog in a trailer. Some like to stay at hotels, some like to camp. Some trips are planned to the day; others ride as it comes. But gear and planning count less than attitude. As one cycle tourist I met said, "all that matters is if you have a smile on your face." (Incidentally, he rose at sunrise, biked until 3 PM, then spent a few hours at the neighborhood bar.) In short, take any bike-tour advice you receive with a grain of salt. The only way to figure out what works for you is to try it out.

Since my first bike tour in 1999, I've logged just shy of 30,000 miles of "loaded" bike touring—touring where I've carried my camping gear. That includes the 21,000 miles of this trip plus a week-long trip in Alaska, a trip down the Pacific Coast, a month-long ride in Eastern Europe, and a trip across the U.S. with my dad after graduating from my master's program—not to mention numerous weekend excursions. With all this pedaling, I've learned what details make for the perfect bike trip—for me.

THE BIKE

On longer, international trips, durability is more important than speed, since the bike breaking down in a remote area could set your journey back a week or more. On longer trips I prefer a steel touring bike. Though steel is heavy, it is less prone to damage and can be fixed via welding. I like twenty-six-inch mountain bike wheels, as they are more commonly found throughout the world than the slightly larger road bike wheels—which is important when you need a spare tire. For the same reason I make sure the bike's rims can take the more common Shrader valves. (If your bike is set up for Presta valves, you can have a bike shop enlarge the hole.)

On shorter trips closer to home, you can get away with a lighter bike. But on any touring bike it is very important to have a strong rear wheel, preferably with thirty-six spokes. If you have panniers, this is where most of your weight will be. Note that replacing a broken spoke on the rear wheel is an enormous pain. (Always carry extra spokes!)

It is extremely important that the bike have low-enough gearing—you'll want all the help you can get when climbing up a steep hill with weight. My smallest chain ring in the front has twenty-two teeth, and the largest in the rear has thirty-four, which is about the lowest gearing you can get.

I also recommend a high-quality pannier rack, preferably made of steel. I've seen many cheap aluminum racks break.

The most important feature of a bike, though, is comfort—nothing matters more. Nearly every cyclist I know has a different preference for a saddle, although the most popular among long-distance cyclists are expensive leather saddles, such as those made by Brooks. I used a nose-less saddle (a Hobson Easyseat), which was easier on my rear but put more weight on my hands. To compensate, I prefer to ride as upright as possible, as such is easier on the neck and hands.

PANNIERS VERSUS TRAILER

When traveling light, the best system is one or two panniers—or, better yet, a single stuff sack on a rear rack, which is lighter and more aerodynamic. When carrying larger loads, you can use either four panniers or a trailer. At first consideration one might think the trailer would add more weight, but sometimes it's only slightly heavier—given the weight of the racks and panniers, and the fact that the bike needs to be stronger to carry more weight. And while a trailer does add some rolling resistance, air resistance is usually more important, and one-wheeled trailers are generally more aerodynamic than four panniers. On my trip, though, I chose to use four panniers because it made the bike easier to carry upstairs and bring into hotel rooms or stores. Also, it fit more easily into pickup trucks, boats, and planes.

GEAR IN GENERAL

The heavier the gear you carry, the heavier your bike has to be. If you want to carry fifty pounds of gear, you will need a bike with extra-strong wheels and a strong rear rack—which means that the bike will have to be stronger and thus heavier to support this weight. Also, the heavier your bike is, the beefier your tires have to be, and the more rolling resistance you will have. Less is more.

STOVE

I usually use an alcohol-burning stove, which I made from two aluminum soda cans. (An Internet search for "alcohol can stove" will generate plenty of guidelines.) It weighs almost nothing. Also, having alcohol on hand is great for killing germs and disinfecting cuts. The only challenge is finding the right alcohol: make sure it is more than 90 percent pure, preferably more than 95 percent. I had the most luck at hardware stores and pharmacies, especially in Latin America—less so in Eastern Europe. Note that these stoves don't always work in cold weather or at higher altitudes, so they wouldn't be the best choice for some trips.

SLEEPING BAG & PAD

A high-quality down sleeping bag can compress to almost nothing. Do take care to not get the bag wet, though this is usually only a challenge in exceptionally wet climates. I like using a thin three-quarter-length inflatable pad underneath. Be sure to bring a patch kit if you opt for something inflatable.

TENT

Choosing your tent wisely can save a lot of weight. I strongly recommend Tarptents (www.tarptent.com), as they are single-walled and well-ventilated. My two-pound Tarptent Rainbow had enough room for me and all of my gear, was sufficiently ventilated for hot nights in the tropics, and kept out heavy rainstorms. I did have to reseal the seams halfway through my journey, but I probably would have had to do that with any tent.

WATER BLADDER

I carry a six- to ten-liter water bladder (I prefer MSR Dromedary bags). I like the independence of not needing to rely on a faucet or stream near my campsite, so a ten-liter bladder is one of my most useful camping items, as it can provide water for dinner and breakfast. I generally fill it around 20 to 30 minutes before looking for a campsite.

EXPANDABLE STUFF SACK

I like to carry an extra waterproof stuff sack; this comes in very handy if I need to ride for a few days without resupplying. I reserve the space atop my rear rack for those occasional extra items.

TOOLS

I generally carry the basic set of tools: Allen wrenches, small crescent wrench, chain tool, extra chain links, patch kit, spoke wrench, and three extra spokes. I also always carry zip ties, duct tape, and extra bolts, just in case. But note: the tools are less important than the mechanic. The best advice is to get lots of experience fixing bikes. Inevitably, though, something will break that you won't be able to fix; at those times you'll have to either rig up something or hitchhike. Be prepared for either scenario.

CLOTHES

In addition to one or two sets of street clothes, I usually carry two pairs of biking shorts and two jerseys. My arm and leg warmers are incredibly useful, and take up little space. I often travel with rain jackets and rain pants that are merely water-resistant rather than waterproof, mostly because these pack down smaller than their GORE-TEX equivalents do. But also, when it rains my goal is not to stay dry, but to stay warm. In that vein, synthetics are much better than cotton—they dry more quickly and keep you warmer when wet.

SHOES

If you use clip-in pedals, make sure you can walk comfortably in them. I prefer to use bike sandals (Shimano makes sandals that can take cleats), so I don't have to wash socks as frequently. (I wear socks and rain covers over my feet in colder or rainier climates.) For much of Latin America I traveled with Chaco sandals and flat pedals; it was great having a single pair of shoes.

PANNIERS

I much prefer waterproof panniers over non-waterproof, as I dislike having to put on a rain cover every time clouds threaten rain. On my trip, though, I had two waterproof panniers and two non-waterproof ones. One advantage to the non-waterproof pannier is it has better air circulation; this was great for storing wet items like my still-damp cooking pot. In addition to the panniers, I also like a handlebar bag, which is a handy stash for a camera and snacks.

FOOD

One of the hardest things to learn is how to eat enough food. Many people don't eat enough on their first tours and spend hours grumpy and frustrated, unaware they're suffering from low blood sugar. So, my advice: eat early and often. Essentially, eat before you're hungry, and don't go for more than two hours without eating something. Each person, of course, is different, and you'll have to find your own eating strategy. Just be sure to consider it a strategy—you burn a lot of calories while riding.

CAMPING

Obviously, learn what the rules are about "free" camping in the places you travel (although, to be honest, I've often not done this). I try to make sure no one sees me when I leave the road to camp, and then I pitch the tent where it is unlikely to be found. Also, be smart. While you might be less safe camping as an individual than with a group, you're also less likely to be found.

If you're traveling in a country where you don't speak the language, at least learn how to say "tent, one night" in that region's tongue. That phrase, plus some sign language, are usually all I need to easily find an overnight spot. And, when asking to camp behind someone's house, be sure you have all the food and water you need—it's best not to ask for anything more than a spot for the night.

Of course, many people will invite you into their homes.

STAYING AT PEOPLE'S HOMES

People are often extremely generous in opening their home to a stranger, perhaps because they're both envious of your adventure and sympathetic to your exhaustion. For the most part don't be shy about accepting what you're offered—few hosts grant their hospitality reluctantly, and in some situations or cultures a refusal might be seen as rude. So, be ready to share stories of your travels in return. And be a good houseguest. This means always being self-sufficient, supplying your own food and water. Also, don't overstay your welcome. I always try to leave on a high note—while my host is still enjoying the visit. I am especially careful if I have to stay anywhere longer than two nights. And be sure to get the person's address to send a thank-you note at trip's end.

FIRE STATIONS

Throughout Latin America, many fire stations may offer an empty bunk for the night. I imagine that someday liability concerns will put an end to this generosity—the reason this plan doesn't often work in the U.S. Of course, offer your hosting fire station the same courtesy you would when staying at someone's home.

———

When Bill Bradlee was deciding whether or not to join me biking across the U.S., someone told him: "If you want to hate this country, read the newspaper every day. If you want to love this country, ride a bicycle across it." Nothing could be truer. The news depicts the world as a scary place—filled with crime, disasters, and other perils. The world you experience from a bike, though, the "real world," is very different, and it is absolutely amazing—full of smiling, generous people who care about the place they call home.

So, take the time to travel. Get on a bike, get out there, and explore the world at ten miles an hour. You will love it more, and it will give you hope.

REFERENCES

A note on sources for the chapter-opener statistics on population, GDP, and CO_2 pollution from fossil fuels—

For California: per capita GDP figures (2012), in 2013 U.S. dollars, from the U.S. Bureau of Economic Analysis; population figures (2012) from the U.S. Census Bureau; per capita CO_2 emissions figures (2010) from the U.S. Census Bureau. Note that, unlike the other countries, this figure does not include cement production, which has only a small effect on the total number—around 5 percent, on average. Data for all other chapters are from the World Bank's World Development Indicators (http://databank.worldbank.org): per capita GDP figures (2012) expressed in 2013 U.S. dollars; population figures (2012); CO_2 emissions figures (2010) from burning fossil fuels and the production of cement.

CHAPTER 1: CALIFORNIA

One-third of the world's population . . . live on two dollars a day or less. Global income comparisons are tricky to make, and "one-third" is a very rough figure. According to one 2005 study by the World Bank, out of 6.5 billion people, 2.6 billion, or almost 40 percent, live on two dollars a day or less. This percentage is fortunately decreasing, largely due to rising incomes in China. See Chen, Shaohua, and Martin Ravallion. "The Developing World Is Poorer than We Thought, But No Less Successful in the Fight Against Poverty." SSRN Scholarly Paper. Rochester, NY: Social Science Research Network, August 1, 2008. http://papers.ssrn.com/abstract=1259575.

As for what those in wealthy countries live on, in the U.S., the median household income as of the time of my journey (2005) was just over $50,000 per year. With an average household size of about 2.6, that means a person living in a median household has about $50,000/2.6/365 = $53 per day. From U.S. Census Bureau. www.census.gov.

The basics of the school presentation concerning the increase in CO_2 in the atmosphere and impacts of climate change is drawn largely from the 2007 IPCC report. See Pachauri, Rajendra K. *Climate Change 2007: Synthesis Report. Contribution of Working Groups I, II, and III to the Fourth Assessment Report* (2008). Cambridge: Cambridge University Press.

California's Central Valley . . . produces over half of the nation's fruits and vegetables. For . . . almonds, apricots, raisins, grapes, olives, pistachios, and walnuts, the valley is responsible for over 90 percent of the nation's harvest. From the California Farm Bureau Federation. Accessed November 10, 2013. www.cfbf.com.

In California . . . a combination of warmer temperatures, less rain, and less snowpack in the Sierra Nevada will likely reduce the yields of every major crop. See Hayhoe, Katharine, Daniel Cayan, Christopher B. Field, Peter C. Frumhoff, Edwin P. Maurer, Norman L. Miller, Susanne C. Moser, et al. "Emissions Pathways, Climate Change, and Impacts on California." Proceedings of the National Academy of Sciences of the United States of America 101, no. 34 (August 24, 2004): 12422–12427. doi:10.1073/pnas.0404500101.

A longer, updated list of papers on climate impacts in California, as well as interactive maps showing changes in temperature, snowpack, rainfall, and wildfires, can also be found at www.cal-adapt.org, a website developed by the California Energy Commission and other key California institutions. See also "2009 California Climate Adaptation Strategy. A Report to the Governor of the State of California in Response to Executive Order S-13-2008." California Natural Resources Agency, 2009. www.energy .ca.gov/2009publications/CNRA-1000-2009-027/CNRA-1000-2009-027-F.PDF.

Mist along the Pacific coast . . . has been decreasing due to warmer temperatures. See Johnstone, James A., and Todd E. Dawson. "Climatic Context and Ecological Implications of Summer Fog Decline in the Coast Redwood Region." Proceedings of the National Academy of Sciences (February 16, 2010). doi:10.1073/pnas.0915062107.

Marine species were already moving north in response to warming temperatures. See Barry, J. P., C. H. Baxter, R. D. Sagarin, and S. E. Gilman. "Climate-Related, Long-Term Faunal Changes in a California Rocky Intertidal Community." *Science* 267, no. 5198 (February 3, 1995): 672–675. doi:10.1126/science.267.5198.672.

The fuel needed to fly a plane full of passengers across the country is equivalent to what would be used if all the passengers drove the same distance, two people per car. The most fuel-efficient U.S. airplanes can fly about 70 miles per gallon per person—which is roughly equivalent to two passengers driving the same distance in a car that gets 35 miles per gallon. See Zeinali, Mazyar, Daniel Rutherford, Irene Kwan, and Anastasia Kharina. "U.S. Domestic Airline Fuel Efficiency Ranking, 2010" (2013). http://trid.trb.org/view. aspx?id=1262316.

This comparison, though, doesn't consider that the contrails of jets may greatly increase the impact of flying (perhaps more than double), meaning that one passenger's individual contribution to the emissions of a cross-country flight might be more than if that passenger drove the same distance alone. See Burkhardt, Ulrike, and Bernd Kärcher. "Global Radiative Forcing from Contrail Cirrus." *Nature Climate Change* 1, no. 1 (April 2011): 54–58. doi:10.1038/nclimate1068.

CHAPTER 2: BAJA CALIFORNIA

Global warming will certainly increase the number of wildfires. See Westerling, A. L., H. G. Hidalgo, D. R. Cayan, and T. W. Swetnam. "Warming and Earlier Spring Increase Western U.S. Forest Wildfire Activity." *Science* 313, no. 5789 (August 18, 2006): 940–943. doi:10.1126/science.1128834.

Most climate models predict that the deserts of Mexico and the southwestern U.S. will become even drier and hotter. See Seager, Richard, Mingfang Ting, Isaac Held, Yochanan Kushnir, Jian Lu, Gabriel Vecchi, Huei-Ping Huang, et al. "Model Projections of an Imminent Transition to a More Arid Climate in Southwestern North America." *Science* 316, no. 5828 (May 25, 2007): 1181–1184. doi:10.1126/science.1139601.

The northeastern shore of the Baja peninsula gets less than four inches of rain a year, while Ensenada, seventy miles away on the western shore, gets about fifteen. See Holmgren, Camille A., Julio L. Betancourt, and Kate A. Rylander. "Vegetation History along the Eastern, Desert Escarpment of the Sierra San Pedro Mártir, Baja California, Mexico." *Quaternary Research* 75, no. 3 (May 2011): 647–657. doi:10.1016/j.yqres.2011.01.008.

The United States is one of the few nations on Earth that didn't ratify the Kyoto Protocol. The others include Afghanistan, Andorra, Brunei, Chad, Iraq, Palestinian National Authority, Sahrawi Arab Democratic Republic (SADR), San Marino, Somalia, Taiwan, Vatican City, and Zimbabwe.

The United States is responsible for nearly 30 percent of the carbon dioxide pollution currently in the atmosphere . . . the developed world is responsible for nearly three-quarters. Note that this is historic—as developing countries catch up, that percentage is dropping, but only slowly. See Baumert, Kevin, Timothy Herzog, and Jonathan Pershing. *Navigating the Numbers: Greenhouse Gases and International Climate Change Agreements.* World Resources Institute, 2005. http://isbndb.com/d/book/navigating_the_numbers.html. Another source, the Carbon Dioxide Information Analysis Center, is updated annually; http://cdiac.ornl.gov.

CHAPTER 3: CENTRAL MEXICO

Over as little time as a few decades, average temperatures in a given region can change quickly, as much as 10 degrees. See Alley, R. B. "Abrupt Climate Change." *Science* 299, no. 5615 (March 28, 2003): 2005–2010. doi:10.1126/science.1081056.

Climate during the past ten thousand years . . . has been far more stable than in the past few hundred thousand years. See Schmidt, M. W. and J. E. Hertzberg. "Abrupt Climate Change During the Last Ice Age." *Nature Education Knowledge,* no. 10 (2011): 11.

If the climate changes such that it is too dry for those evergreens, one likely scenario would be that the forest is gradually replaced by grassland. As for that possible transition: though we can't be sure, it would likely result from both tree mortality and fire. The most extensive modeling studies have been done on the Amazon; see Malhi, Yadvinder, Luiz E. O. C. Aragão, David Galbraith, Chris Huntingford, Rosie Fisher, Przemyslaw Zelazowski, Stephen Sitch, Carol McSweeney, and Patrick Meir. "Exploring the Likelihood and Mechanism of a Climate-Change-Induced Dieback of the Amazon Rainforest." Proceedings of the National Academy of Sciences 106, no. 49 (December 8, 2009): 20610–20615. doi:10.1073/pnas.0804619106.

Global warming would alter rain patterns . . . more rain would fall during the colder months in sufficient quantity to kill wintering monarchs. See Oberhauser, Karen, and A. Townsend Peterson. "Modeling Current and Future Potential Wintering Distributions of Eastern North American Monarch Butterflies." *Proceedings of the National Academy of Sciences* 100, no. 24 (November 25, 2003): 14063–14068. doi:10.1073/pnas.2331584100.

In this century nearly 15 percent of species may go extinct. See Maclean, Ilya M. D., and Robert J. Wilson. "Recent Ecological Responses to Climate Change Support Predictions of High Extinction Risk." *Proceedings of the National Academy of Sciences* 108, no. 30 (July 26, 2011): 12337–12342. doi:10.1073/pnas.1017352108.

A 2-degree Celsius (3.5-degree Fahrenheit) increase in global temperatures . . . would threaten 20 to 30 percent of all known species on the planet. See Thomas, Chris D., Alison Cameron, Rhys E. Green, Michel Bakkenes, Linda J. Beaumont, Yvonne C. Collingham, Barend F. N. Erasmus, et al. "Extinction Risk from Climate Change." *Nature* 427, no. 6970 (January 8, 2004): 145–148. doi:10.1038/nature02121.

Even a warming of just a few degrees will decrease yields in tropical nations. David Lobell, a professor at Stanford, has published a number of articles on this topic. One example is Lobell, David B., J. Ivan Ortiz-Monasterio, Gregory P. Asner, Pamela A. Matson, Rosamond L. Naylor, and Walter P. Falcon. "Analysis of Wheat Yield and Climatic Trends in Mexico." *Field Crops Research* 94, no. 2–3 (November 15, 2005): 250–256. doi:10.1016/j.fcr.2005.01.007.

By the end of the century the average summer temperatures in Mexico could be warmer than the warmest summers ever recorded in the region. See Battisti, David S., and Rosamond L. Naylor. "Historical Warnings of Future Food Insecurity with Unprecedented Seasonal Heat." *Science* 323, no. 5911 (January 9, 2009): 240–244. doi:10.1126/science.1164363.

[Mexico's] average crop yields could decrease by 25 percent. See Cline, William R. "Global Warming and Agriculture: Impact Estimates by Country" (2007). http://socionet.ru/publication.xml?h=repec:iie:ppress:4037.

In just a few decades, almost all of Mexico will have a climate considered "severe drought" by today's standards. See Dai, Aiguo. "Drought under Global Warming: A Review." *Wiley Interdisciplinary Reviews: Climate Change* 2, no. 1 (2011): 45–65. doi:10.1002/wcc.81.

CHAPTER 4: MEXICO CITY

The description of the Aztec capital Tenochtitlán derives from Mann, Charles C. *1491: New Revelations of the Americas Before Columbus.* Random House LLC, 2005, 130–132.

Smog forms more readily in warmer climates. See Hogrefe, Christian. "Air Quality: Emissions versus Climate Change." *Nature Geoscience* 5, no. 10 (October 2012): 685–686. doi:10.1038/ngeo1591.

Mexico's total carbon dioxide pollution is one-tenth the amount of that of the United States. See Martinez, Julia, and Adrián Fernández Bremauntz. *Cambio climático: una visión desde México.* Mexico City: Instituto Nacional de Ecología, 2004, 110–111.

Deforestation accounts for roughly 20 percent of greenhouse gas pollution worldwide. According to the 2007 IPCC report, deforestation accounted for approximately 17 percent of global annual emissions. See Pachauri, Rajendra K. *Climate Change 2007: Synthesis Report. Contribution of Working Groups I, II, and III to the Fourth Assessment Report* (2008). Cambridge: Cambridge University Press, 37.

Today the forests of Europe and North America . . . are expanding. See Kauppi, Pekka

E., Jesse H. Ausubel, Jingyun Fang, Alexander S. Mather, Roger A. Sedjo, and Paul E. Waggoner. "Returning Forests Analyzed with the Forest Identity." *Proceedings of the National Academy of Sciences* 103, no. 46 (November 14, 2006): 17574–17579. doi:10.1073/pnas.0608343103.

The average Mexican uses about one-sixth the amount of electricity as the average American, and produces less than a quarter as much carbon dioxide from fossil fuels. From 2011 figures, World Bank Open Data, http://data.worldbank.org.

CHAPTER 5: SOUTHERN MEXICO

Mexico has a highly unequal distribution of wealth; most of the country is poor, but as much as one-fifth of the population is more wealthy than the average United States' citizen. Based on data from www.gapminder.org.

If [the ice sheets of Greenland and Antarctica] melt, sea levels will rise two hundred feet—the height of a twenty-story building. . . . Thirty-five million years ago, the global average temperature was about 9 degrees Fahrenheit warmer, sea levels were 200 feet higher, and there was no ice at either pole. See Jansen, E., J. Overpeck, K. R. Briffa, J.-C. Duplessy, F. Joos, V. Masson-Delmotte, D. Olago, et al., 2007: "Palaeoclimate." In: *Climate Change 2007: The Physical Science Basis. Contribution of Working Group I to the Fourth Assessment Report of the Intergovernmental Panel on Climate Change* (Solomon, S., D. Qin, M. Manning, Z. Chen, M. Marquis, K. B. Averyt, M. Tignor and H. L. Miller, eds.). Cambridge: Cambridge University Press, 2007, 440–442. Available at www.ipcc.ch.

If sea level were to increase by two feet this century—which is very likely. See Horton, Benjamin P., Stefan Rahmstorf, Simon E. Engelhart, and Andrew C. Kemp. "Expert Assessment of Sea-Level Rise by AD 2100 and AD 2300." *Quaternary Science Reviews* 84 (January 15, 2014): 1–6. doi:10.1016/j.quascirev.2013.11.002.

Mangroves' roots provide the habitat for shrimp and are the spawning grounds for many types of fish that later travel out to sea. See Sasekumar A, V. C. Chong, M. U. Leh, and R. D'Cruz. 1992. "Mangroves as a Habitat for Fish and Prawns." *Hydrobiologia* 247:195–207.

If the Earth warms . . . quickly, sea level rise could overwhelm the mangroves' ability to [adapt]. See Valiela, Ivan, Jennifer L. Bowen, and Joanna K. York. "Mangrove Forests: One of the World's Threatened Major Tropical Environments." *Bioscience* 51 (2001): 807–815.

If the two billion biomass users instead heated their food and warmed their homes with more efficient stoves, their health would improve and global warming would be slowed. See Revkin, Andrew. "Soot in the Greenhouse, and Kitchen." *The New York Times.* March 26, 2008. See also Rosenthal, Elisabeth. "Third-World Stove Soot Is Target in Climate Fight." *The New York Times.* April 15, 2009.

CHAPTER 6: GUATEMALA

The discussion with Mark Brenner derives from both our actual conversation and additional research for verification. Brenner, Mark, personal communication, September 2010. See Hodell, David, Flavio Anselmetti, Mark Brenner, and Daniel Ariztegui. "The

Lake Petén Itzá Scientific Drilling Project." *Scientific Drilling,* no. 3 (September 2006). See also Anselmetti, Flavio S., David A. Hodell, Daniel Ariztegui, Mark Brenner, and Michael F. Rosenmeier. "Quantification of Soil Erosion Rates Related to Ancient Maya Deforestation." *Geology* 35, no. 10 (October 1, 2007): 915–918. doi:10.1130/ G23834A.1.

The Maya used [the concept of zero] hundreds of years in advance of the Old World. See Mann, Charles C. *1491: New Revelations of the Americas before Columbus.* New York: Vintage, 2006.

I tried to imagine what Tikal must have looked like mid-abandonment. Description of population density, and how the lands surrounding Tikal were cleared for agriculture, derive from the following: Diamond, Jared M. *Collapse: How Societies Choose to Fail or Succeed.* New York: Penguin Books, 2011. And Haug, Gerald H., Detlef Günther, Larry C. Peterson, Daniel M. Sigman, Konrad A. Hughen, and Beat Aeschlimann. "Climate and the Collapse of Maya Civilization." *Science* 299, no. 5613 (March 14, 2003): 1731–1735. doi:10.1126/science.1080444. And Hodell, David A., Mark Brenner, and Jason H. Curtis. "Terminal Classic Drought in the Northern Maya Lowlands Inferred from Multiple Sediment Cores in Lake Chichancanab (Mexico)." *Quaternary Science Reviews* 24, no. 12–13 (July 2005): 1413–1427. doi:10.1016/j .quascirev.2004.10.013.

[That] the Intertropical Convergence Zone (ITCZ) . . . didn't migrate as far north . . . may have been caused by natural variability in the Earth's climate, or it may have been because Maya deforestation affected the local weather. See Pringle, Heather. "A New Look at the Mayas' End." *Science* 324, no. 5926 (April 24, 2009): 454–456. doi:10.1126 /science.324_454.

Climate models predict that this region of northern Central America will become far drier in the twenty-first century, potentially even drier than during the times of the ancient Maya. See Christensen, J. H., B. Hewitson, A. Busuioc, A. Chen, X. Gao, I. Held, R. Jones, et al., 2007: "Regional Climate Projections." In: *Climate Change 2007: The Physical Science Basis. Contribution of Working Group I to the Fourth Assessment Report of the Intergovernmental Panel on Climate Change* (Solomon, S., D. Qin, M. Manning, Z. Chen, M. Marquis, K. B. Averyt, M. Tignor, et al., eds.). Cambridge: Cambridge University Press, 2007, 895.

CHAPTER 7: BELIZE

Coral reefs are the world's second-most biologically diverse ecosystem, trailing only tropical rainforests in diversity of life. Though they cover only 0.1 percent of the ocean's floor, they are home to one-third of all species of aquatic plants and animals, including at least 4,000 species of fish. See Stone, Richard. "A World Without Corals?" *Science* 316, no. 5825 (May 4, 2007): 678–681. doi:10.1126/science.316.5825.678.

Over the past few decades, the oceans have absorbed about a third of our carbon dioxide pollution, thus keeping the levels of carbon dioxide in the atmosphere lower than they would have been otherwise. But, in the process, the oceans have become more acidic—by 0.1 pH. See Doney, Scott C., Victoria J. Fabry, Richard A. Feely, and Joan A. Kleypas. "Ocean Acidification: The Other CO_2 Problem." *Marine Science* 1 (2009).

"We're creating conditions in the ocean that we haven't seen since the dinosaurs went extinct. By mid-century, reefs might not be able to grow their shells." Caldeira, Ken. Personal communication, 2005.

CHAPTER 8: HONDURAS

Computer climate models suggest hurricanes will become more powerful in a warmer world. See Knutson, Thomas R., and Robert E. Tuleya. "Impact of CO_2-Induced Warming on Simulated Hurricane Intensity and Precipitation: Sensitivity to the Choice of Climate Model and Convective Parameterization." *Journal of Climate* 17, no. 18 (September 2004): 3477–3495. doi:10.1175/1520-0442(2004)017<3477:IOCWOS>2.0.CO;2. Also, an MIT study showed that, as a result of warming ocean temperatures, the energy of hurricanes worldwide has increased by 60 percent over the past three decades. See Emanuel, Kerry. "Increasing Destructiveness of Tropical Cyclones over the Past 30 Years." *Nature* 436, no. 7051 (August 4, 2005): 686–688. doi:10.1038/nature03906.

Both climate models and scientific observation suggest that . . . increased rainfall will . . . be concentrated in the same number of heavier storms. As the Earth has warmed over the past fifty years, in some places storm sizes have increased even though total rainfall decreased— meaning that the rain fell in fewer, heavier storms. See U.S. Global Change Research Program. *Global Climate Change Impacts in the United States.* Cambridge: Cambridge University Press, 2009.

I felt guilty because humans have already emitted perhaps half of the carbon dioxide the atmosphere can absorb if we are to keep the Earth's warming within a safe level, less than 2 degrees Celsius (3.5 degrees Fahrenheit). See Allen, Myles R., David J. Frame, Chris Huntingford, Chris D. Jones, Jason A. Lowe, Malte Meinshausen, and Nicolai Meinshausen. "Warming Caused by Cumulative Carbon Emissions towards the Trillionth Tonne." *Nature* 458, no. 7242 (April 30, 2009): 1163–1166. doi:10.1038/nature08019.

CHAPTER 9: EL SALVADOR

Remesas, or remittances, is the money workers in the U.S. send home to their families in Latin America. . . . [S]uch "income" is now over 15 percent of the country's GDP. See Castillo-Ponce, Ramon A., Victor Hugo Torres-Preciado, and Jose Luis Manzanares-Rivera. "Macroeconomic Determinants of Remittances for a Dollarized Economy: The Case of El Salvador." *Journal of Economic Studies* 38, no. 5 (September 27, 2011): 562–576. doi:10.1108/01443581111161823.

Twenty percent of Central America's forests had been cut down in the fifteen years before my journey; only sparsely populated Belize has avoided major deforestation. See "Neotropical Realm: Environmental Profile." Accessed November 24, 2013. http://rainforests.mongabay.com/20neotropical.htm.

[Deforestation] also accounts for over half of the region's greenhouse gas pollution. See Harris, Nancy, Sandra Brown, Stephen C. Hagen, Alessandro Baccini, and Richard Houghton. "Progress Toward a Consensus on Carbon Emissions from Tropical Deforestation." Policy Brief, Meridian Institute (2012).

In the 1980s and 1990s, twenty-nine sub-Saharan countries had civil wars. . . . [T]he best predictor of whether a country would go to war was drought. See Miguel, Edward, Shanker

Satyanath, and Ernest Sergenti. "Economic Shocks and Civil Conflict: An Instrumental Variables Approach." *Journal of Political Economy* (2004): 725–753.

The conflict in Darfur arguably may have been influenced by lack of rain. See Faris, Stephan. "The Real Roots of Darfur." *The Atlantic,* April 2007. www.theatlantic.com/magazine/archive/2007/04/the-real-roots-of-darfur/305701.

India is already building a six-foot concrete wall along the border with Bangladesh to keep out immigrants. See Campbell, Kurt M., Jay Gulledge, J. R. McNeill, John Podesta, Peter Ogden, Leon Fuerth, R. James Woolsey, Alexander T. J. Lennon, Julianne Smith, and Richard Weitz. *The Age of Consequences: The Foreign Policy and National Security Implications of Global Climate Change.* Washington, DC: Center for a New American Security, 2007, 5.

CHAPTER 10: NICARAGUA

The level of scientific consensus on climate change is almost unprecedented. Nearly every scientific organization . . . have all endorsed statements that humans are causing the Earth to warm. For a summary of the scientific consensus see www.skepticalscience.com.

CHAPTER 11: COSTA RICA

Statistics comparing Costa Rica with its neighbors derive from U.N. Human Development Reports. "Indices & Data: Human Development Reports (HDR): United Nations Development Programme (UNDP)." Accessed November 24, 2013. http://hdr.undp.org/en/statistics.

The Environmental Kuznets Curve . . . idea is that, as a poor country's economy grows, the nation initially degrades its environment; but, once the economy grows past a certain point, the country's citizens demand better environmental protection, and environmental degradation slows or even reverses. See "The China Syndrome and the Environmental Kuznets Curve—Energy and the Environment—AEI." Accessed November 24, 2013. www.aei.org/article/energy-and-the-environment/the-china-syndrome-and-the-environmental-kuznets-curve. See also Culas, Richard J. "Deforestation and the Environmental Kuznets Curve: An Institutional Perspective." *Ecological Economics* 61, no. 2–3 (March 1, 2007): 429–437. doi:10.1016/j.ecolecon.2006.03.014.

Some evidence suggests that Latin America is already slowing its deforestation, as deforestation rates across the region are lower than they were a decade earlier. Measuring deforestation is challenging. The Food and Agriculture Organization (FAO) found deforestation in Latin America was slightly lower from 2000 to 2010 than it was from 1990 to 2000, and deforestation in Brazil, the largest deforester, has dropped dramatically. See *Global Forest Resources Assessment 2010: Main Report—FAO Forestry Paper 163.* Food and Agriculture Organization of the United Nations. Rome: 2010, xvi. www.fao.org/docrep/013/i1757e/i1757e.pdf. See also Hansen, M. C., P. V. Potapov, R. Moore, M. Hancher, S. A. Turubanova, A. Tyukavina, D. Thau, et al. "High-Resolution Global Maps of 21st-Century Forest Cover Change." *Science* 342, no. 6160 (November 15, 2013): 850–853. doi: 10.1126/science.1244693.

Dr. Pounds showed that the region became drier because of warmer ocean temperatures; the warm oceans caused clouds to form at higher altitudes, reducing the mist and moisture in the cloud forest. See Pounds, J. Alan, Michael P. L. Fogden, and John H. Campbell.

"Biological Response to Climate Change on a Tropical Mountain." *Nature* 398, no. 6728 (April 15, 1999): 611–615. doi:10.1038/19297.

While some more recent research disputes that climate change is responsible for the golden toad's demise, there is general agreement that a changing climate makes such extinctions more likely. See Anchukaitis, Kevin J., and Michael N. Evans. "Tropical Cloud Forest Climate Variability and the Demise of the Monteverde Golden Toad." *Proceedings of the National Academy of Sciences* 107, no. 11 (March 16, 2010): 5036–5040. doi:10.1073/pnas.0908572107.

In one single genus of toads, Atelopus, . . . many species have gone extinct over the past twenty years. . . . [A]lmost all . . . have followed unusually warm years. See Pounds, J. Alan, Martín R. Bustamante, Luis A. Coloma, Jamie A. Consuegra, Michael P. L. Fogden, Pru N. Foster, Enrique La Marca, et al. "Widespread Amphibian Extinctions from Epidemic Disease Driven by Global Warming." *Nature* 439, no. 7073 (January 12, 2006): 161–167. doi:10.1038/nature04246.

CHAPTER 12: PANAMA

Per mile it's fifteen times more efficient to ship by boat than by truck. See Parajuli, Ashis, Luis Ferreira, and Jonathan M. Bunker. "Freight Modal Energy Efficiency: A Comparison Model." In *Faculty of Built Environment and Engineering.* University of South Australia, Adelaide, 2003. http://eprints.qut.edu.au/307.

[Ships] produce only a small fraction of carbon dioxide pollution from transportation, and are responsible for only a little more than 1 percent of all greenhouse gas pollution. See McCollum, David L., Gregory Gould, and David L. Greene. *Greenhouse Gas Emissions from Aviation and Marine Transportation: Mitigation Potential and Policies.* Institute of Transportation Studies, Working Paper Series. Institute of Transportation Studies, UC Davis, January 1, 2010. http://econpapers.repec.org/paper/cdlitsdav/qt5nz642qb.htm.

A study by researchers at Carnegie Mellon University showed that only about 10 percent of the carbon emissions from food derives from its transportation. The majority comes from growing and processing the food. See Weber, Christopher L., and H. Scott Matthews. "Food-Miles and the Relative Climate Impacts of Food Choices in the United States." *Environmental Science & Technology* 42, no. 10 (May 1, 2008): 3508–3513. doi:10.1021/es702969f.

Because sheep are raised much more efficiently in New Zealand than in England, British citizens' carbon footprint would be smaller if they ate meat imported from the far side of the Earth rather than from the local farm. See Saunders, Caroline, Andrew Barber, and Greg Taylor. "Food Miles—Comparative Energy: Emissions Performance of New Zealand's Agriculture Industry" (July 2006). http://researcharchive.lincoln.ac.nz/handle/10182/125.

World trade has exploded in recent years, doubling in the decade before my journey. From World Trade Organization website, www.wto.org.

CHAPTER 13: COLOMBIA

Designing . . . communities to be more conducive to bicycling. . . . [I]n Bogotá, personal vehicle use has decreased, bicycle use and walking have increased, and people like the city more. See Montezuma, Ricardo. "The Transformation of Bogotá, Colombia, 1995–2000:

Investing in Citizenship and Urban Mobility." *Global Urban Development* 1, no. 1. Facing the Environmental Challenge (2005).

Something else remarkable had happened in Colombia in the decade before my visit: the country had reduced its carbon dioxide pollution. Vergara, Walter. Lead Engineer, Latin America Environment Department, World Bank. Personal Communication. 2008.

The TransMilenio and bikeways also had a serious effect, perhaps decreasing Bogotá's pollution by over half a million tons of carbon dioxide a year, and cutting Bogotá's total pollution by a few percentage points. See Kahn Ribeiro, S., S. Kobayashi, M. Beuthe, J. Gasca, D. Greene, D. S. Lee, Y. Muromachi, et al., 2007: "Transport and its Infrastructure." In *Climate Change 2007: Mitigation. Contribution of Working Group III to the Fourth Assessment Report of the Intergovernmental Panel on Climate Change* (B. Metz, O. R. Davidson, P. R. Bosch, R. Dave, L. A. Meyer, eds.). Cambridge: Cambridge University Press, 350.

The average citizen of Los Angeles produces about five tons [of CO_2 per year from transportation]. See Kennedy, Christopher, Julia Steinberger, Barrie Gasson, Yvonne Hansen, Timothy Hillman, Miroslav Havránek, Diane Pataki, Aumnad Phdungsilp, Anu Ramaswami, and Gara Villalba Mendez. "Greenhouse Gas Emissions from Global Cities." *Environmental Science & Technology* 43, no. 19 (October 1, 2009): 7297–7302. doi:10.1021/es900213p.

The average citizen of Copenhagen is responsible for less than one and a half tons [of CO_2 per year through from transportation]. See Madsen, Jørgen Lund. Initiatives Against Climate Changes and Need for Future Actions in Copenhagen: A presentation given at the Fourth Municipal Leaders Summit on Climate Change, in Montreal, Canada. 2005.

In . . . 2009 all of the world's transportation was responsible for less than seven billion tons of CO_2. See International Energy Agency. "CO_2 Emissions from Fossil Fuel Combustion: Highlights." 2011. www.iea.org/publications/freepublications/publication/co2highlights.pdf.

If the Earth warms by 4.5 to 5.5 degrees Fahrenheit, we'd lose 97 percent of the páramo. See Buytaert, W. "The Impact of Climate Change on the Water Supply of the Andean Highlands." In *Geophysical Research Abstracts,* vol. 8 (2006): 3058.

The páramo alone is home to some 2,000 species of plants and animals, many of which are found only in Colombia. See Spehn, Eva M., Maximo Liberman, and Christian Korner. *Land Use Change and Mountain Biodiversity.* Boca Raton, FL: CRC Press, 2006, 112.

CHAPTER 14: BOLIVARIAN REPUBLIC OF VENEZUELA

Throughout Latin America, about ten million people might have to move their homes if sea levels rise just three feet. See Anthoff, D., R. J. Nicholls, R. S. J. Tol, and A. T. Vafeidis. "Global and Regional Exposure to Large Rises in Sea-Level: A Sensitivity Analysis." Monograph, 2006. www.tyndall.ac.uk/content/global-and-regional-exposure-large-rises-sea-level-sensitivity-analysis-work-was-prepared-st.

If we were to use CCS on all coal power plants in the world, we would have to annually pump around twenty billion tons of carbon dioxide underground. See Chu, Steven. "Carbon Capture and Sequestration." *Science* 325, no. 5948 (September 25, 2009): 1599. doi:10.1126/science.1181637.

In eastern Venezuela we have about a trillion barrels of oil underground. Note that not all of this is "recoverable"—but with advances in technology, it very well might be. See James, K. H. "The Venezuelan Hydrocarbon Habitat." *Geological Society, London, Special Publications* 50, no. 1 (January 1, 1990): 9–35. doi:10.1144/GSL.SP.1990.050.01.02.

As oil extraction doesn't require many workers, it does not employ a large percentage of the population; as a result, an already poor distribution of wealth gets skewed even further. See Sachs, Jeffrey D., and Andrew M. Warner. "The Curse of Natural Resources." *European Economic Review* 45, no. 4–6 (May 2001): 827–838. doi:10.1016/S0014-2921(01)00125-8.

Average Venezuelans were poorer during my trip than they'd been three decades earlier. In 2006, during my journey, per capita GDP in Venezuela was $5,401 USD; in 1976 it was $6,383 USD (measured in 2000 dollars). From World Bank Open Data, http://data.worldbank.org.

The few oil-rich countries that have escaped the so-called "curse," such as Dubai, Mexico, or Indonesia, have done so not by increasing government spending, but by diversifying their economy away from oil and investing in manufacturing or service industries. See Brunnschweiler, C. N., and E. H. Bulte. "Linking Natural Resources to Slow Growth and More Conflict." *Science* 320, no. 5876 (May 2, 2008): 616–617. doi:10.1126/science.1154539. See also Schubert, Samuel R. "Revisiting the Oil Curse: Are Oil Rich Nations Really Doomed to Autocracy and Inequality?" MPRA Paper, August 1, 2006. http://mpra.ub.uni-muenchen.de/10109.

Between when Chávez took power in 1999 and when I biked across the country nearly a decade later, 40 percent of industrial companies were shuttered. See Gould, Jens Erik. "With Oil's Cash, Venezuelans Consume." *The New York Times,* June 8, 2006. www.nytimes.com/2006/06/08/business/worldbusiness/08spend.html.

Caracas may have been the most dangerous city I visited during my entire journey. Its murder rate, at about six thousand per year, was then nearly nine times that of Bogotá. Caracas's murder rate in 2009 was 2 per 1,000 people. See Romero, Simon. "Venezuela, More Deadly Than Iraq, Debates Why." *The New York Times,* August 22, 2010. www.nytimes.com/2010/08/23/world/americas/23venez.html.

Since mosquitoes survive better in warmer weather, they can spread more disease [such as malaria and dengue fever] in warmer climates. See Patz, Jonathan A., Diarmid Campbell-Lendrum, Tracey Holloway, and Jonathan A. Foley. "Impact of Regional Climate Change on Human Health." *Nature* 438, no. 7066 (November 17, 2005): 310–317. doi:10.1038/nature04188.

Because of climate change . . . malaria already appears to be creeping up the mountains. See Epstein, Paul R., Henry F. Diaz, Scott Elias, Georg Grabherr, Nicholas E. Graham, Willem J. M. Martens, Ellen Mosley-Thompson, and Joel Susskind. "Biological and Physical Signs of Climate Change: Focus on Mosquito-Borne Diseases." *Bulletin of the American Meteorological Society* 79, no. 3 (March 1998): 409–417. doi:10.1175/1520-0477(1998)079<0409:BAPSOC>2.0.CO;2.

During the summer of 2003 about thirty-five thousand people, mostly in France, died prematurely from an unusually strong heat wave. Confalonieri, U., B. Menne, R. Akhtar, K. L. Ebi, M. Hauengue, R. S. Kovats, B. Revich, et al., 2007: Human Health. *Climate Change 2007: Impacts, Adaptation and Vulnerability. Contribution of Working Group II to the Fourth Assessment Report of the Intergovernmental Panel on Climate Change,* M. L. Parry, O. F. Canziani, J. P. Palutikof, P. J. van der Linden, and C. E. Hanson, eds., Cambridge: Cambridge University Press, 2007, 397.

A surprising number of . . . diseases are more common in warmer weather. For more on the health impacts of climate change see: "The World Health Report 2002: Reducing Risks, Promoting Healthy Life." *WHO.* Accessed November 25, 2013. www.who.int /whr/2002/en.

While Chávez may have been to blame for Venezuela's precarious political and economic state, so is the so-called "curse of oil." Juan Pablo Pérez Alfonso, Venezuela's Minister of Mines and Hydrocarbons of Venezuela, in the 1970s said that oil is "The excrement of the devil." See "Economics Focus: The Devil's Excrement." *The Economist,* May 22, 2003. www.economist.com/node/1795921.

CHAPTER 15: BRAZIL

Livestock are responsible for nearly one-fifth of global greenhouse gas pollution—nearly the amount of pollution from all cars, planes, trains, and ships combined. See Steinfeld, Henning, Pierre Gerber, T. Wassenaar, V. Castel, Mauricio Rosales, and C. de Haan. "Livestock's Long Shadow" (2006). www.fao.org/docrep/010/a0701e/a0701e00.htm.

The average American or Western European eats a thousand calories a day from animal sources, while the average person in the world consumes fewer than five hundred. "ESS: FAO Statistical Yearbook 2005–2006." Accessed November 25, 2013. www.fao.org /economic/la-direccion-de-estadistica/publicaciones-estudios/statistical-yearbook /fao-statistical-yearbook-2005-2006/es.

The biodiversity of the Amazon jungle is among the world's highest. . . . it's estimated that one out of every ten species on Earth lives within this basin. See Wilson, Edward O. *The Future of Life.* New York: Vintage Books, 2003.

Some climate models predict that global warming will dry out the Amazon Basin, and . . . [that] the majority of the Amazon would give way to savannah or desert. See Cox, P. M., R. A. Betts, M. Collins, P. P. Harris, C. Huntingford, and C. D. Jones. "Amazonian Forest Dieback under Climate-Carbon Cycle Projections for the 21st Century." *Theoretical and Applied Climatology* 78, no. 1–3 (June 1, 2004): 137–156. doi:10.1007 /s00704-004-0049-4.

Although a few more recent studies have suggested [the Amazon becoming savannah] isn't highly likely, it is still possible. See Cox, Peter M., David Pearson, Ben B. Booth, Pierre Friedlingstein, Chris Huntingford, Chris D. Jones, and Catherine M. Luke. "Sensitivity of Tropical Carbon to Climate Change Constrained by Carbon Dioxide Variability." *Nature* 494, no. 7437 (February 21, 2013): 341–344. doi:10.1038/nature11882.

A little bit of warming will result in so-called "positive feedback" loops that will accelerate warming. See: Field, Christopher B., and Michael R. Raupach. *The Global Carbon Cycle: Integrating Humans, Climate, and the Natural World.* Washington, DC: Island Press, 2004.

An enormous amount of organic material is locked away in the tundra of Siberia and northern North America. Warmer temperatures would allow soil microbes to digest this organic material, which would emit huge amounts of methane and carbon dioxide to the atmosphere.

See Schuur, Edward A. G., James Bockheim, Josep G. Canadell, Eugenie Euskirchen, Christopher B. Field, Sergey V. Goryachkin, Stefan Hagemann, et al. "Vulnerability of Permafrost Carbon to Climate Change: Implications for the Global Carbon Cycle." *BioScience* 58, no. 8 (2008): 701–714.

Some archeologists suggest that Indian populations greatly influenced the forests, selectively promoting fruit trees, and that a number of the trees I saw in the forest were actually "domesticated." Mann, Charles C. *1491: New Revelations of the Americas Before Columbus.* New York: Vintage, 2006, 337–344.

Paper on deforestation in the Amazon . . . : "Here is the Brazilian forest in 1992, and in 2002. The dark gray is deforestation." See See Correia, Francis Wagner S., Regina Célia Dos S. Alvalá, Antonio O. Manzi. "A GCM Simulation of Impact of Land Cover Changes in the Amazonia on Regional Climate." *Proceedings of 8 ICSHMO.* Fozdo Iguaçu, Brazil: INPE—National Institute For Space Research (April 24–28, 2006): 873-878.

In the half decade since my bike trip, something remarkable has happened in Brazil: the deforestation rate has dropped dramatically. See Malingreau, J. P., H. D. Eva, and E. E. de Miranda. "Brazilian Amazon: A Significant Five Year Drop in Deforestation Rates but Figures Are on the Rise Again." *AMBIO* 41, no. 3 (May 1, 2012): 309–314. doi:10.1007/s13280-011-0196-7.

CHAPTER 16: PERU I

According to climate models, [it's likely] that the [Amazon's] seasonality will become stronger, with a wetter rainy season and a drier dry season. See Marengo, J. A., C. A. Nobre, J. Tomasella, M. F. Cardoso, and M. D. Oyama. "Hydro-Climatic and Ecological Behaviour of the Drought of Amazonia in 2005." *Philosophical Transactions of the Royal Society B: Biological Sciences* 363, no. 1498 (May 27, 2008): 1773–1778. doi:10.1098/rstb.2007.0015.

During the 2005 [Amazon] drought, many settlements were stranded, as the river's flow was too low for ships to reach them. The Brazilian Armed Forces spent three months airlifting supplies to towns, delivering two hundred tons of food and thirty tons of medicine. See Rohter, Larry. "Record Drought Cripples Life Along the Amazon." *The New York Times,* December 11, 2005. www.nytimes.com/2005/12/11/international/americas/11amazon.html.

In this region along the Andes, as much as a quarter of the forests have been cut down, mostly by farmers who had migrated over the mountains to Peru's frontier. See Portuguez, Hubert, Patricia Huerta. *Mapa de Deforestación de la Amazonia Peruana—2000.* Instituciones Participantes Instituto Nacional de Recursos Naturales Consejo Nacional del Ambiente. Government of Peru. 2005. http://geoservidor.minam.gob.pe/geoservidor/archivos /memoria/DEFORESTACION_Parte1.pdf.

CHAPTER 17: PERU II

A quarter of the ice [in the Cordillera Blanca] had melted in the past thirty years [since the mid-1970s], dramatically changing the landscape. See Racoviteanu, A. E., Yves Arnaud, M. W. Williams, and J. Ordonez. "Decadal Changes in Glacier Parameters in the Cordillera Blanca, Peru, Derived from Remote Sensing." *Journal of Glaciology* 54, no. 186 (2008): 499–510. doi:10.3189/002214308785836922.

Huaráz may see ten inches of rain in February, but less than one inch in June and July. See Bryan G. Mark, Geoffrey O. Seltzer. "Evaluation of Recent Glacier Recession in the Cordillera Blanca, Peru (AD 1962–1999): Spatial Distribution of Mass Loss and Climatic Forcing." *Quaternary Science Reviews* (2005): 2265–2280. doi:10.1016/j.quascirev.2005.01.003.

Losing the glaciers might decrease only one-third of the water flow. See Baraer, Michel, Bryan G. Mark, Jeffrey M. McKenzie, Thomas Condom, Jeffrey Bury, Kyung-In Huh, Cesar Portocarrero, Jesús Gómez, and Sarah Rathay. "Glacier Recession and Water Resources in Peru's Cordillera Blanca." *Journal of Glaciology* 58, no. 207 (February 1, 2012): 134–150. doi:10.3189/2012JoG11J186.

Weather stations high in the Andes have warmed much faster than at sea level. See Bradley, Raymond S., Mathias Vuille, Henry F. Diaz, and Walter Vergara. "Threats to Water Supplies in the Tropical Andes." *Science* 312, no. 5781 (June 23, 2006): 1755–1756. doi:10.1126/science.1128087.

One model suggests that by 2050, 40 to 60 percent of the remaining glaciers will be gone, and water flowing in the Río Santa during the dry season could be greatly reduced. See Juen, Irmgard, Georg Kaser, and Christian Georges. "Modelling Observed and Future Runoff from a Glacierized Tropical Catchment (Cordillera Blanca, Perú)." *Global and Planetary Change* 59, no. 1–4 (October 2007): 37–48. doi:10.1016/j.gloplacha.2006.11.038.

In 1950 less than 40 percent of Latin Americans lived in cities. See Nash, Nathaniel. "Squalid Slums Grow as People Flood Latin America's Cities." *The New York Times.* Accessed November 25, 2013. www.nytimes.com/1992/10/11/world/squalid-slums-grow-as-people-flood-latin-america-s-cities.html.

Today that figure [Latin Americans who live in cities] is almost 80 percent. See Paranagua, Paulo A. "Latin America Struggles to Cope with Record Urban Growth." *The Guardian,* September 11, 2012. www.theguardian.com/world/2012/sep/11/latin-america-urbanisation-city-growth.

In some ways, the urban poor have a higher quality of life than their rural counterparts. Statistics on the urban poor in Latin America derive from Fay, Marianne. *The Urban Poor in Latin America.* Washington, DC: World Bank Publications, 2005.

In the 1980s, a period of economic stagnation and the terrorism of the Shining Path, the city of Lima annually gained as many as four hundred thousand rural villagers trying to escape violence and poverty. See Nash, Nathaniel. "Squalid Slums Grow as People Flood Latin America's Cities." *The New York Times.* Accessed November 25, 2013. www.nytimes.com/1992/10/11/world/squalid-slums-grow-as-people-flood-latin-america-s-cities.html.

In the past decade Peru has doubled the amount of coal it burns. See U.S. Energy Information Administration, www.eia.gov. For a graph of Peru's coal consumption, go to www.indexmundi.com/energy.aspx?country=pe&product=coal&graph=consumption.

The average Peruvian currently uses one-fifteenth the electricity the average U.S. citizen uses. From World Bank Open Data, http://data.worldbank.org.

One-sixth of the world's population gets water from rivers fed by snow and glaciers. . . . In just a few decades, rivers around the world will be far more variable, and many rivers, including those in Peru, will see less flow during the dry months. See Barnett, T. P., J. C. Adam, and D. P. Lettenmaier. "Potential Impacts of a Warming Climate on Water Availability in Snow-Dominated Regions." *Nature* 438, no. 7066 (November 17, 2005): 303–309. doi:10.1038/nature04141.

CHAPTER 18: BOLIVIA

The city actually has a high-altitude ski resort, but snow no longer reaches the bottom of the resort's only lift. See Romero, Simon. "Bolivia's Only Ski Resort Is Facing a Snowless Future." *The New York Times,* February 2, 2007. www.nytimes.com/2007/02/02/world /americas/02bolivia.html.

Twenty-five percent of Bolivians live on less than two dollars a day. From World Bank Open Data, http://data.worldbank.org.

A similar percentage [25 percent] are malnourished. See FAO, WFP, and IFAD. 2012. *The State of Food Insecurity in the World 2012: Economic Growth Is Necessary But Not Sufficient to Accelerate Reduction of Hunger and Malnutrition.* Rome, FAO.

During the past ice age, . . . much of the Altiplano was covered by a series of lakes, the largest of which, Lake Minchin, was the size of modern-day Lake Michigan. See Bills, Bruce G., Shanaka L. de Silva, Donald R. Currey, Robert S. Emenger, Karl D. Lillquist, Andrea Donnellan, and Bruce Worden. "Hydro-Isostatic Deflection and Tectonic Tilting in the Central Andes: Initial Results of a GPS Survey of Lake Minchin Shorelines." *Geophysical Research Letters* 21, no. 4 (1994): 293–296. doi:10.1029/93GL03544.

The Great Salt Lake in Utah rises and falls depending on rain levels, such that a rise in the lake during the 1980s threatened to inundate Salt Lake City's airport. See Williams, Terry Tempest. *Refuge: An Unnatural History of Family and Place.* New York: Vintage Books, 1992.

In the past century, . . . the level [of the Caspian Sea] fluctuated by twenty feet, enough to flood oil rigs along its northern coast—ecological disasters from just the past few decades. See Rodionov, S. N. *Global and Regional Climate Interaction: The Caspian Sea Experience.* Norwell, MA: Kluwer Academic, 1994.

During an El Niño year . . . warmer waters slightly reverse the normal weather patterns, making the coast wetter than normal and bringing drought to the Amazon and Altiplano. See Garreaud, René, Mathias Vuille, and Amy C. Clement. "The Climate of the Altiplano: Observed Current Conditions and Mechanisms of Past Changes." *Palaeogeography, Palaeoclimatology, Palaeoecology* 194, no. 1–3 (May 15, 2003): 5–22. doi:10.1016 /S0031-0182(03)00269-4. See also Knüsel, S., S. Brütsch, K. A. Henderson, A. S. Palmer, and M. Schwikowski. "ENSO Signals of the Twentieth Century in an Ice Core from Nevado Illimani, Bolivia." Journal of Geophysical Research: Atmospheres 110, no. D1 (2005). doi:10.1029/2004JD005420.

In 2000, during a La Niña year, heavy rain caused Lake Titicaca to rise by five feet, and two hundred thousand people found their homes under water. See "Peru Floods Situation Report 21 Mar 2001." *ReliefWeb,* March 21, 2001. http://reliefweb.int/report/peru /peru-floods-situation-report-21-mar-2001.

Some models predict a stronger El Niño, others a stronger La Niña. See Collins, Matthew, and The CMIP Modelling Groups (BMRC [Australia], CCC [Canada], CCSR/NIES

[Japan], CERFACS [France], CSIRO [Australia], MPI [Germany], GFDL [U.S.], GISS [U.S.], IAP [China], INM [Russia], LMD [France], MRI [Japan], NCAR [U.S.], NRL [U.S.], Hadley Centre [U.K.] and YNU [South Korea]). "El Niño– or La Niña–like Climate Change?" *Climate Dynamics* 24, no. 1 (December 9, 2004): 89–104. doi:10.1007/s00382-004-0478-x.

According to some, but not all, paleoclimatologists, the Pacific was in a permanent El Niño state during [the Pliocene Epoch]. See Wara, Michael W., Ana Christina Ravelo, and Margaret L. Delaney. "Permanent El Niño–like Conditions During the Pliocene Warm Period." *Science* 309, no. 5735 (July 29, 2005): 758–761. doi:10.1126/science.1112596.

Half of the world's known supplies [of lithium] are in the salt flats of [Bolivia, Argentina, and Chile]. See Quinones, Manuel. "Global Lithium Deposits Enough to Meet Electric Car Demands—Report." *The New York Times,* July 28, 2011. www.nytimes.com/gwire/2011/07/28/28greenwire-global-lithium-deposits-enough-to-meet-electri-67078.html.

Today, there are one billion cars in the world, and it is estimated that in the next few decades human society will add another billion. See "Transport Outlook 2011: Meeting the Needs of 9 Billion People." OECD-International Transport Forum, 2011.

According to some estimates, there's enough lithium in just Bolivia and northern Argentina and Chile to provide batteries for more than four billion cars. See Smith, Michael, and Matthew Craze. "Lithium for 4.8 Billion Electric Cars Lets Bolivia Upset Market." *Bloomberg News,* December 7, 2009. www.bloomberg.com/apps/news?pid=newsarchi ve&sid=aVqbD6T3XJeM.

CHAPTER 19: NORTHERN CHILE

Ernesto "Che" Guevara, crossing Latin America by motorcycle, had also met mine workers in Calama. For more on Che's encounter with the mine workers and the debated importance it had on his life, see: Castañeda, Jorge G. *Compañero: The Life and Death of Che Guevara.* New York: Vintage Books, 1998, 47.

The average income in Latin America is almost three times higher today than it was fifty years ago, the infant mortality rate has dropped by a factor of five, and the average Latin American born today can expect to live almost twenty years longer than his or her counterpart born in 1950. See CEPALSTAT Databases and Statistical Publications: Economic Commission for Latin America and the Caribbean. Accessed November 25, 2013. http://estadisticas.cepal.org/cepalstat/WEB_CEPALSTAT/Portada.asp?idioma=i.

CHAPTER 20: NORTHERN ARGENTINA

The subtropics are already slightly drier, and the higher latitudes are receiving, on average, more rain than they did fifty or a hundred years earlier. See Zhang, Xuebin, Francis W. Zwiers, Gabriele C. Hegerl, F. Hugo Lambert, Nathan P. Gillett, Susan Solomon, Peter A. Stott, and Toru Nozawa. "Detection of Human Influence on Twentieth-Century Precipitation Trends." *Nature* 448, no. 7152 (July 26, 2007): 461–465. doi:10.1038/nature06025.

CHAPTER 21: CENTRAL CHILE

Climate models suggest that both central Chile and California will likely see decreased precipitation in a warmer world. See "Chile Faces Climate Change Challenge." *BBC,* May 23, 2009. http://news.bbc.co.uk/2/hi/8058080.stm. See also Magrin, G., C. Gay García, D. Cruz Choque, J. C. Giménez, A. R. Moreno, G. J. Nagy, C. Nobre, et al., 2007: "Latin America." In *Climate Change 2007: Impacts, Adaptation and Vulnerability. Contribution of Working Group II to the Fourth Assessment Report of the Intergovernmental Panel on Climate Change,* M. L. Parry, O. F. Canziani, J. P. Palutikof, P. J. van der Linden, and C. E. Hanson, eds., Cambridge: Cambridge University Press, 598–599.

In the process of seizing control of [Chile], Pinochet had imprisoned perhaps eighty thousand Chileans, tortured as many as thirty thousand—even women and children—and killed two to three thousand. From *Report of the Chilean National Commission on Truth and Reconciliation.* Notre Dame, Indiana: University of Notre Dame Press, 1993. www .usip.org/sites/.../truth_commissions/...Report/Chile90-Report.pdf. Also from *The National Commission on Political Imprisonment and Torture Report* (June 1, 2005). www .archivochile.com/Derechos_humanos/com_valech/Informe_complementario.pdf (in Spanish).

In some countries—such as Venezuela, Bolivia, Nicaragua, and Peru—the average citizen was poorer during my journey than they'd have been thirty years earlier. From World Bank Open Data, http://data.worldbank.org.

Since 1990, [Chile's] annual fossil fuel pollution has increased by about 50 percent, from about 2.6 tons to nearly 4 tons. . . . Per capita electricity use has more than tripled since 1980. From World Bank Open Data, http://data.worldbank.org.

In the next few decades, almost all the growth in carbon dioxide pollution will come from developing nations such as the ones I biked through. The International Energy Agency estimates that the growth in developing countries will increase the global demand for energy by about 50 percent in the next twenty-five years. See *World Energy Outlook 2012.* International Energy Agency. www.worldenergyoutlook.org/publications/weo-2012.

Soon after my trip ended the Chilean congress passed a bill pledging that by 2014 Chile would produce at least 5 percent of its energy from non-hydroelectric renewable energy, increasing that amount to at least 10 percent by 2024. See "Chile's Congress Approves Renewable Energy Law." Reuters, March 6, 2008. http://uk.reuters.com/article/2008/03/06 /environment-energy-chile-dc-idUKN0563031120080306.

CHAPTER 22: NORTHERN PATAGONIA

According to the International Energy Agency, [South America] could nearly quadruple the electricity it generates from hydropower. See International Energy Agency. *Technology Roadmap: Hydropower.* 2012. www.iea.org/publications/freepublications/publication/2012 _Hydropower_Roadmap.pdf.

South America already gets nearly 70 percent of its electricity from such dams. See International Energy Agency: Statistics. Accessed March 10, 2013. www.iea.org/statistics.

Our reshaping of the natural world dates back to the first humans, who were so successful at hunting they forever altered the ecosystems by killing off creatures like the mastodons and woolly mammoths. Across the world, before 10,000 BC, 70 percent of the world's animals bigger than 100 pounds were hunted to extinction. The Americas used to be inhabited by numerous species of giant sloths, some of which approached the size of African elephants. Many natural history museums offer panoramas of giant mammals, such as huge species of deer or wooly mammoths, all of which were wiped out between the last ice age and today. Studies and modeling suggest the best explanation for these extinctions is the expertise of human hunters. See Alroy, John. "A Multispecies Overkill Simulation of the End-Pleistocene Megafaunal Mass Extinction." *Science* 292, no. 5523 (June 8, 2001): 1893–1896. doi:10.1126/science.1059342. See also Doughty, Christopher E. "The Development of Agriculture in the Americas: An Ecological Perspective." Ecosphere 1, no. 6 (December 1, 2010): art21. doi:10.1890 /ES10-00098.1.

Some scientists estimate that between 20 and 40 percent of all plant production on the planet is used by humans—from agriculture to ranch land to forestry. See Vitousek, Peter M., Paul R. Ehrlich, Anne H. Ehrlich, and Pamela A. Matson. "Human Appropriation of the Products of Photosynthesis." *BioScience* 36, no. 6 (June 1986): 368–373. doi:10.2307/1310258. But, since estimating quantities of plant production appropriated by humans can be tricky, note that another paper, which considered just land use of plant production, estimated that humans use 10–50 percent of the world's plant production. See Rojstaczer, Stuart, Shannon M. Sterling, and Nathan J. Moore. "Human Appropriation of Photosynthesis Products." *Science* 294, no. 5551 (December 21, 2001): 2549–2552. doi:10.1126/science.1064375.

We may have to bring another 2.5 to 5 billion acres of land into cultivation . . . in order to feed our civilization. See Sachs, Jeffrey, Roseline Remans, Sean Smukler, Leigh Winowiecki, Sandy J. Andelman, Kenneth G. Cassman, David Castle, et al. "Monitoring the World's Agriculture." *Nature* 466, no. 7306 (July 29, 2010): 558–560. doi:10.1038/466558a. See also "How to Feed a Hungry World." Nature 466, no. 7306 (July 29, 2010): 531–532. doi:10.1038/466531a.

A survey of the 270 glacial tongues that lead off [the Campo de Hielo Sur ice sheet] indicates that all but two are rapidly retreating. See Glasser, N. F., S. Harrison, K. N. Jansson, K. Anderson, and A. Cowley. "Global Sea-Level Contribution from the Patagonian Icefields since the Little Ice Age Maximum." *Nature Geoscience* 4, no. 5 (May 2011): 303–307. doi:10.1038/ngeo1122.

CHAPTER 23: TIERRA DEL FUEGO

One study estimated that the installation of turbines across Argentina could produce, on average, over 4.5 terawatts of electricity. See Lu, Xi, Michael B. McElroy, and Juha Kiviluoma. "Global Potential for Wind-Generated Electricity." *Proceedings of the National Academy of Sciences* (June 22, 2009). doi:10.1073/pnas.0904101106.

While the Arctic is expected to warm much more than the rest of the planet, the southern tip of South America is not. See Pachauri, Rajendra K. *Climate Change 2007: Synthesis Report. Contribution of Working Groups I, II, and III to the Fourth Assessment Report* (2008). Cambridge: Cambridge University Press, 46.

EPILOGUE: RIDE FOR CLIMATE USA

Farmers in several states . . . complained about drought greatly reducing their yields of wheat or corn. See Burke, Marshall, and Kyle Emerick. *Adaptation to Climate Change: Evidence from US Agriculture.* SSRN Scholarly Paper. Rochester, NY: Social Science Research Network, September 11, 2012. http://papers.ssrn.com/abstract=2144928.

Wyoming's Bighorn Mountains have lost miles and miles of ponderosa pine forest to bark beetles—infestations linked to warmer temperatures and drier weather. See Raffa, Kenneth F., Erinn N. Powell, and Philip A. Townsend. "Temperature-Driven Range Expansion of an Irruptive Insect Heightened by Weakly Coevolved Plant Defenses." *Proceedings of the National Academy of Sciences* 110, no. 6 (February 5, 2013): 2193–2198. doi:10.1073/pnas.1216666110. See also Chapman, Teresa B., Thomas T. Veblen, and Tania Schoennagel. "Spatiotemporal Patterns of Mountain Pine Beetle Activity in the Southern Rocky Mountains." *Ecology* 93, no. 10 (May 8, 2012): 2175–2185. doi:10.1890/11-1055.1.

About four million Americans live so close to the ocean that the projected rise of sea level, perhaps by three feet this century, will endanger their homes and communities. See Anthoff, D., R. J. Nicholls, R. S. J. Tol, and A. T. Vafeidis. "Global and Regional Exposure to Large Rises in Sea-Level: A Sensitivity Analysis." Monograph, 2006. www.tyndall.ac.uk/sites/default/files/wp96_0.pdf.

When given a list of twenty different issues—ranging from deficit reduction to Medicare to tax policy—U.S. citizens consistently ranked global warming as the least important. See "Twelve Years of the Public's Top Priorities." *Pew Research Center for the People and the Press.* Accessed December 19, 2013. www.people-press.org/interactives/top-priorities.

Made in the USA
San Bernardino, CA
08 March 2014